AF306341

FIREARM BLUEING AND BROWNING

R. H. ANGIER

Foreword by Ned Schwing

STACKPOLE
BOOKS

Published by
STACKPOLE BOOKS
4501 Forbes Boulevard, Suite 200
Lanham, Maryland 20706
www.stackpolebooks.com

Distributed by NATIONAL BOOK NETWORK

ISBN 978-0-8117-0326-0

Cataloging-in-Publication Data is on file with the Library of Congress

TABLE OF CONTENTS

Part 1. General Division

Part 2. Special Division

FOREWORD

Have you ever wondered why guns are blued? Have you ever wondered how it is done? I'm not talking about blueing a firearm with a solution out of a bottle, but about how firearms are blued by military armories, manufacturers, and small custom builders. *Firearm Blueing and Browning*, by R.H. Angier, describes not only these processes, but also how gun metal was darkened in the sixteenth, seventeenth, and eighteenth centuries. The historic formulae and processes are discussed in sufficient detail so that an individual can, with the right equipment and time, duplicate these older procedures.

For those readers who have the equipment to blue or brown their guns, this book provides an indispensable source of information. Angier discusses the various processes of degreasing, derusting, rusting, and darkening. He also provides crucial instructions on the preparation of blueing and browning solutions. The solution's chemicals are described with their chemical compounds, effects on metals, uses, and results.

The bulk of *Firearm Blueing and Browning* is devoted to what Angier refers to as "working instructions." It is here that the reader is given detailed formulas for a number of blueing and browning procedures. For example, browning with common salt, one of the oldest metal finishes, is detailed. The classic browning process for eighteenth century British army weapons, "Antimony brown," is also discussed. The list goes on with dozens of blueing and browning formulas described.

For those of us who do not finish our own metals, we now have the necessary information to give to a professional antique firearm restorer. Angier's book is invaluable to those who want to restore a firearm to its original finish using the right chemicals and time-honored techniques.

Ned Schwing

PREFACE

This book originated in the urge experienced sooner or later
by almost every gun-lover, to undertake adjustments or modifications to, or reconditionings of his own firearms.

In the literature of metal colouring and gunsmithing, the subject of browning and blueing is rather cursorily treated: in
particular, accurate data in respect of chemicals, the results of
experience in particular cases, "kinks" for overcoming troubles,
and criticism of compositions and methods, are mostly con-
spicuous by their absence.

The present work is specially addressed to the individual gun
owner, the gunsmith and maker of small precision work: the
larger Armouries disposing of skilled technical experts and all
available aids, while their mass-production manufacture permits,
indeed compels, the installation of equipment quite beyond the
reach of the individual or general gunsmith.

Chemical colouring methods are here alone described, which
can be carried out by anybody without special plant or the aid
of electricity, excluding such as are only suitable for rougher
work. For the benefit of the individual gun-owner, it is important to insist, right here, with all possible emphasis, that
browning of irreproachable quality in no wise depends on completeness of equipment or even on special manual skill, but
above all on patience and the care taken in carrying out the
very simple stages of the process: these pre-assumed, results in
every respect equal to those characterizing the work of the finest
luxury gunmakers are obtainable with the simplest domestic
means.

In compiling the present work, the author's main principle
has been insistance on precision (wherever attainable) in place

of the usual vagueness, uniformity to facilitate comparison, and as far as possible, critical appreciation derived from long experience, of this or that composition or method.

In the same order of ideas, instead of a planless collection of formulae, he offers for the first time a classification of the very numerous compositions, which fall qualitatively into a very few, well-defined series.

The brief but precise definition of the various chemicals, the enumeration of the various trade qualities to be found in different countries, their correct names, and the signification of the often out-of-date and occasionally utterly misleading popular designations sometimes applied to them will often save interested readers of foreign technical literature waste of time and irritation. In this spirit the work is respectfully offered to the private gun-owner and gunsmith.

The author gladly takes the opportunity of tendering his warm thanks to Mr. D. B. Wesson, Vice-President of Smith & Wesson Inc.: to Colonel Townsend Whelen, U. S. Army: to Mr. Harry Brearley (the inventor of stainless steel) of Brown, Bayley's Steel Works, Sheffield, Engl.: to Mr. C. E. Greener (W. W. Greener Ltd., Birmingham, Engl.): to Messrs. A. G. Parker & Co. Ltd., of the same city: to M. M. Auguste Francotte, Liége, Belgium: the Schweizerische Industrie-Gesellschaft, Neuhausen, Switzerland, and all who have kindly afforded him information and assistance for the purpose of this work.

Topsham, Exeter, England.

October, 1936.

PART I

General Division

CHAPTER I

INTRODUCTION

Historical

The production of a dark colour on iron and steel objects, whether for ornamentation or other purposes, reaches far back in the history of metal working. One need only recall the century-old suits of armour, coloured by the action of heat alone, or combined with fats, ox gall, mixtures containing sulphur or antimony glance, or by pickling with acids or other rust-producing substances.

Improved methods followed the development of chemistry, and the oldest browning process for military arms, by means of "butter of antimony" (according to G. Buchner) is described in the *Hanover Magazine* in 1781.

In the literature on the subject the statement is generally found that the browning of gun barrels with this substance originated in England towards the end of the 18th century, which may be quite correct as regards military arms. The browning of barrels is in itself however much older, even in England, and according to information kindly given the author by Mr. C. E. Greener (the well-known Birmingham gunmaker), was in common use for sporting arms about 1720. A report of 1637 in the London Record Office explicitly mentions the "russetting" of barrels under the heading of "Repairs to the Arms of the Trained Bands" (London Militia).

A similar development took place at the same time on the European Continent: guns of the 18th century with blued (temper-blued) and browned barrels are numerous, and of the 17th century occasionally met with, in every larger public or private collection of firearms.

Exactly the same advance took place on the American Continent, and the American backwoodsmen, using the simplest possible means, browned the barrels of their historically famous, home-made rifles, from about the middle of the 18th century onwards.

The very practical innovation was rapidly adopted for private arms, but much more slowly—according to common report, on account of the difficulty of finding men skilled in its application —for military rifles, for which it only became the standard finish towards the latter end of the 19th century (in Prussia-Germany only after 1871).

Nature of Gun-browning

The primary object of "browning"—in the U. S. often but incorrectly described as "blueing"—(Fr. "bronzage," G. "Brünieren," "Bräunen" or "Bronzieren"), gun parts is to get rid of the reflections from polished surfaces which otherwise dazzle the shooter and betray his position to the game or enemy, also to provide a certain degree of rust-protection, and to improve appearance.

In principle, it is nothing more than artificial rusting, effected by special oxidising mixtures or pastes, which when repeatedly applied and brushed off after taking action, produce the desired effect.

As already mentioned, the first process used on a large scale for military arms (in England) produces a coating of warm brown tone, and was hence described as "browning", but in the United States it is more commonly known as "blueing"— probably on account of the more easily produced finish of "temper-blueing" almost universally used for the most characteristic American firearm, the revolver.

At the present day both descriptions are incorrect in 99 cases of 100, a black finish, produced by a modification of the original process, now being practically universal: but the old terms have been respectively retained in the two countries, and must be understood as merely conventional designations, without prejudice to the actual colour of the finish produced.

The author offers no apology for concisely describing the various oxidising mixtures and solutions used as "blues" or "brownes", and for the occasional use of more or less "coined" expressions combining descriptiveness with brevity.

CHAPTER II

THE BLUEING PROCESS (OR BROWNING)

According to its final result, there are two variants of the process, according as the rust is left to itself, or modified by the subsequent action of boiling water or steam: the first giving a brown, the second a black finish.

The rust first formed consists mostly of ferrous hydroxide, which on exposure gradually becomes converted into brown ferric oxide. Subjected to the action of boiling water or steam, the brown oxide is converted into the much darker, so-called "magnetic" oxide (ferro-ferric oxide), the coating thereby mostly becoming black. By suitable additions to the browning solution (and to a lesser degree by after-treatment) the coating can be given a more or less brownish, blueish or black-blueish tone, without modifying the fundamental difference between brown and black colouring.

The present-day preference for black browning is due to its more pleasing appearance and to the considerable saving of work in its production.

The rust produced by the action of a single application of the browne (solution) is mostly of greyish tone, much too light for the end in view, and is deepened by successive applications: in order to secure uniform colouring the loose rust produced by each "rusting" is removed by a wire brush (so-called "scratching"). When dark enough the coating is "fixed" by scalding, then oiling, and sometimes varnishing: this was the original browning process.

This alternation of rusting and "scratching" is the characteristic of the browning process: hollow and other surfaces which cannot be reached by the brush cannot therefore be "browned" in the strict sense, but can be coloured brown or black at will without scratching by one or other of the collective processes described in Chapter IV. G.

Black-browning on a small or medium scale is carried out as follows:

1. Degreasing parts (pickling or matting when desired).
2. Application of browne (browning solution).
3. Rusting (generally at room temperature).
4. Darkening by boiling (less frequently by steaming).
5. Scratching.
    Repeating operations 2 to 5 until the coating is dark enough (mostly 3 to 4 times).
6. Neutralising (exceptionally).
7. Fixing.
    After-lacquering is unnecessary with black-browning (blueing).

The progress of rusting at ordinary temperature varies between hours and days according to weather conditions (temperature and moisture in particular), which usually matters little to the private shooter or small business, but would never do for Government and other large Armouries, regularity being indispensable to mass-production methods. In these, rusting takes place in special cupboards or cells fitted with steam and warm air circulation, thereby not only accelerating rusting, but securing independence of weather conditions, and large, regular output under the most economical conditions. The following is a typical example of large-scale operation.

1. Degreasing (tumbling barrel, boiling in lime-water or soda lye).
2. Application of browning solution, then 10 to 20 minutes in drying oven at 35-40°C. (90-105°F.).
3. After cooling, rusting in steam oven for 20 to 45 minutes at 35-70°C. (95-160°F.).
4. Drying rusted parts in cell in current of dry air.
5. Scratching (after cooling).
    Repeating steps 2 to 5, from 2 to 6 times, generally 3 times.
6. Fixing.

In this way the time taken by the whole process is reduced to about as many hours as it would otherwise take days.

The process will now be described in detail.

Pre-treatment of Parts

Apart from a few exceptions—mentioned hereafter—the appearance of the finished parts corresponds to that of the unbrowned surface: highly-polished surfaces thus retain a brilliant finish, dull ones a correspondingly non-reflecting coating. The mechanical preparation (polishing, etc.) of the pieces thus postulates the necessary attention: the greater the care taken in this respect the better the final result.

1. Degreasing

Absolutely complete degreasing of the pieces is the essential first condition of a satisfactory result, otherwise the metal "takes" the solution with difficulty, incompletely or unevenly, causing a patchy, unpleasing appearance. Neglect in degreasing bears bitter fruit in the sequel, and generally speaking cannot be completely made good in most cases. It is indeed then better to start afresh, (by de-browning as detailed in part 2, chap. VI).

In large-scale manufacture, degreasing is mostly effected by rolling with whiting (Fr. "blanc de Troyes" or "d'Espagne," G. "Schlämmkreide" or "Wienerkalk"), rotten stone (Fr. "pierre pourrie," G. "Kieselgur" or "Tripelerde") alone or mixed with whiting, or with slaked lime, preferably in a tumbling barrel or bowl: or boiled in steam-heated or direct-fired pans with milk of lime, washing soda or (about 5%) caustic potash or soda lye, followed by washing with hot water to remove adhering alkali.

It is to be noted that while alkalies decompose ("saponify," that is, convert into soap) all animal and vegetable fats, the soap resulting from this decomposition emulsifies mineral oil, but does not eliminate it with unconditional certainty. As the use of mineral oils is constantly extending, this point must be borne in mind.

Subject to this reservation, degreasing by tumbling or boiling can be used at choice in moderate scale gunsmithing, but where browning is only occasionally done or on a small scale, the parts are most conveniently degreased by repeated washing with ordinary auto-gasoline or benzol: these bodies dissolve all fats including mineral oils, and themselves evaporate without residue. On account of its great inflammability, caution is necessary in using gasoline: carbon tetrachloride and trichlorethylene (see Chapter III.5, "Chemicals") are incombustible, but on the other hand deleterious and much more expensive, therefore only usable with restrictions.

The elimination of the last traces of grease, especially on mirror-polished surfaces, is often difficult: in such cases, wiping with strong alcohol, 5 to 10% caustic soda lye (and subsequent rinsing with hot water) or with the specially effective but very dangerous carbon disulphide (Chapter III.5), will mostly give the desired result. In desperate cases very slight pickling e. g.

with weak potassium bisulphate or boric acid, can be resorted to, then accepting the slight loss of polish as the lesser evil. (see Chapter V.). Immediate wetting of the parts by water indicates that degreasing is complete.

If the degreased pieces are not to be browned immediately, they must be stored in a completely dry place, or if this is not available, laid for *not too long a time* in boiled (that is, degassed) water containing from 75 to 150 grs. per U. S. quart of purified cream of tartar or slaked lime. Small parts can be kept with advantage in a desiccator (glass vessel with ground cover, sieve and quantity of calcium chloride) which absorbs every trace of moisture.

From this moment onwards the degreased parts must not be touched with the naked hand, as the skin is always slightly fatty: clean rags, cotton gloves, wooden handles, tongs or the like must alone be used. The bore of barrels is carefully oiled and the ends sealed by tightly fitting wooden plugs, which also form convenient handles.

2. Derusting, Denickelling, Pickling and Matting

After degreasing, these operations, if needful or desired, can be carried out: details in Chapters V and VI.

A "matt" (dull) surface and finish are often desirable, e. g. on breechbolts, sights, etc.

3. Applying Browning Solution

Uniform application of the browne to the entire surface of the parts is equally important to secure a pleasing finish, whence the old-time rule, "apply the smallest possible quantity." To this end a pad of cottonwaste, sponge, rag or brush is dipped in the browne, then squeezed nearly dry and applied uniformly to the surfaces.

As a fixed rule, this can only be accepted with a certain reserve: in cases depending on the nature of metal and solution, thus only ascertainable by preliminary trial, more generous application gives decidedly better results: if the action is extremely slow one or more re-applications—allowing complete drying between each—are of distinct advantage.

This procedure for rifle barrels would be far too slow and cumbersome for Armory use: the barrels, sealed at both ends as before mentioned, are dipped singly into the browning solution, then stood up on suitable sinks or grids to drip. Other parts of complicated shape but not requiring the tight closing of hollows or bores, can be similarly treated.

All-too liberal application, and still more emphatically, partial re-application to the same surface when incompletely dry, is to be avoided, as the rust already formed is then partially dissolved, giving rise to crusts, irregular ridges, and finally to an unsightly, patchy coating.

In all this the least neglect in degreasing, or accidental contact with the naked hand, makes itself unpleasantly seen.

4. Rusting

The painted pieces are now left to rust. The rate of rusting depends on the composition of the metal and solution, on the surrounding conditions of temperature and moisture—thus on the weather—and on the physical condition of the metal surfaces (highly polished, matted, hardened, casehardened, partly browned as in reconditioning, etc.), so that no general rule can be stated. That moisture and heat combined promote rusting, is however a bedrock principle.

Natural rusting. Another old-time rule lays down that slowly produced coatings, that is, by weak solutions, many repetitions and "natural" rusting, i. e. at room temperature, are closer in grain and more adherent, therefore more durable, than those formed by rapid methods. This rule is unquestionably based on an easily understood and logical basis—Rome was proverbially not built in a day—but the condition of many weapons collected from the battlefields, browned by the accelerated methods of mass-production, and subjected to the roughest possible usage, suggests a certain reservation.

It must not be forgotten that private shooters' guns receive infinitely more careful handling, which alone would account for the greater durability of their appearance. Further, that this rule dates from the time when browning had not overcome its "teething troubles" and was far from having reached its present-day perfection. The author's conclusion therefore is that the accelerated browning methods of the large Armouries in all countries produce a result completely satisfying all reasonable demands as to appearance and durability.

Accelerated rusting. *Equipment.* Accelerated rusting is universal in mass-production, volume and regularity of output being of primary importance. To this end, as already indicated, special iron ovens or cells are provided, fitted with radiators or other heating appliances, steam supply and warm and cold air circulation, in which operations 2 to 4 (page 6) take

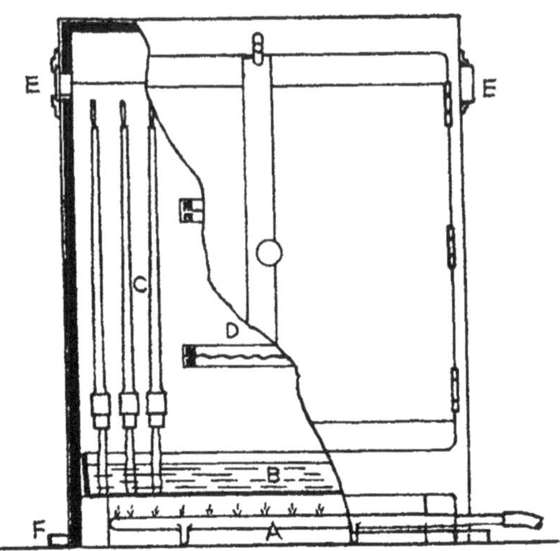

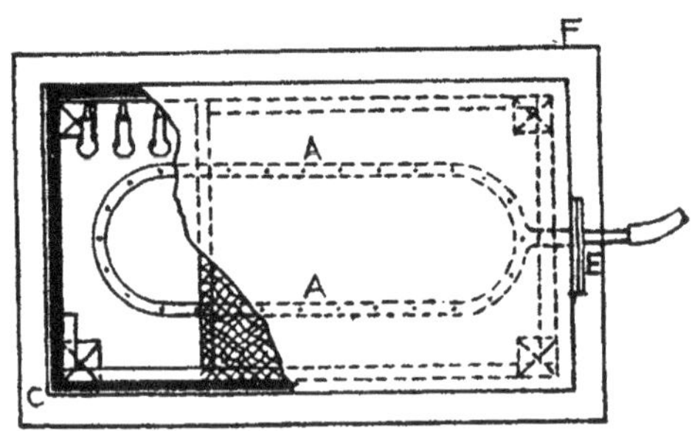

A. *burner.*
B. *water trough.*
C. *barrel compartment.*
D. *grids.*
E. *vapour escape valves.*
F. *tray to catch drip.*

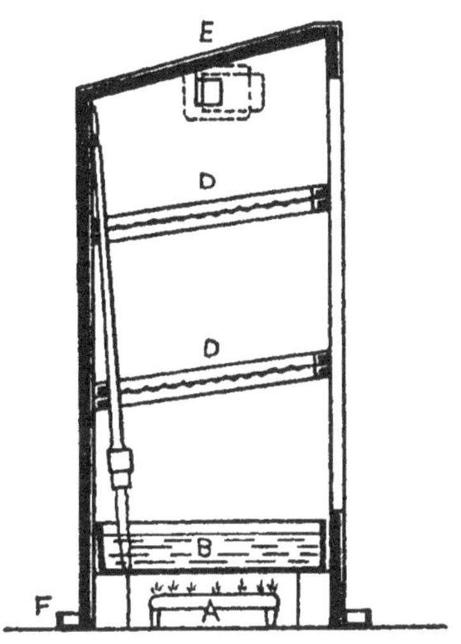

E
D
D
B
A
F

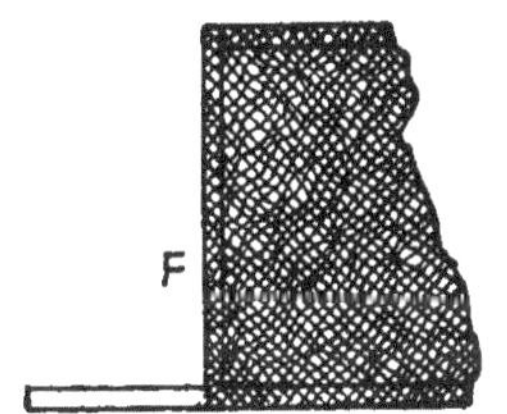

F

place. Barrels are stacked in lots of 20 to 50 on suitable stands, other parts laid on grids or wire netting.

A simpler arrangement suffices for custom gunmakers, consisting of a cupboard of sufficient size, fitted at the bottom with a water trough heated by any convenient means (gas burners, rings or coil, Primus cookstoves, etc.), having a space or compartment to take barrels, and one or more loose grids (wire netting or perforated zinc) for smaller parts: ceiling and grids are sloped to prevent condensate from dripping on the parts below. The annexed figure shows a typical arrangement, which is self-explanatory, and of which, modifications to suit special conditions, suggest themselves: barrels, for instance, can be laid horizontally on suitable supports, but should then be given a half-turn occasionally to ensure uniform action.

The body can be of wood (preferably zinc-lined) or metal (galvanised) as preferred: in the former case the trough should fit easily and the whole stood in a shallow metal tray to catch drip: in the latter, the bottom can be made watertight, and provided with a tap to drain off the contents for periodical flushing. A wooden body requires less heat to maintain a given temperature and keeps it more uniform: a metal one is more durable and not subject to warping.

The temperature usually maintained in the steam oven is about 35 to 40°C. (95-105°F.), less frequently as high as 70°C. (160°F.): the burners are regulated to give the desired temperature and to allow a little steam to escape continually from the valves, thereby proving the "saturated" state of the interior atmosphere.

Where steam is available it advantageously replaces burners and water trough: pressure is quite unnecessary, and excessive steam supply to be avoided, as drip would more or less wash away the applied solution, with the result of patchy rusting.

For smaller parts the private gun-owner can with advantage use a chemist's drying oven containing a water dish, or even more simply, where browning is only occasionally done, by the makeshift composed of two ordinary (degreased!) kitchen bowls or saucepans with a piece of wire netting between and a heavy cloth or other non-conducting cover over the upper one, to keep the temperature inside steadier: with a little practice this makes a very effective substitute.

The parts are "ripe" for scratching when they are seen to be covered with loose rust: they should then be removed from the steam oven, as if left in too long pitting is likely to occur.

On first application and at room temperature "ripeness" is commonly reached after a few hours, but experience is here the only guide, so inspect occasionally. As can be readily understood rusting becomes progressively slower on repeated operations, so that in general and in commercial working more than 3 to 4 passes (processing) are not advantageous. If these do not give a satisfactory result, some other browning solution should be tried.

"Express" browning. Lately, and in the U. S. in particular, speedier methods have become popular, and various trade preparations are advertised at names such as "20-minute blue," often at fancy prices. Although such designations are not to be taken literally, it is nevertheless true that by their means a satisfactory colour can be obtained from start to stop within $1\frac{1}{2}$ to 2 hours (Details in Chapter IV, brownes Ac and Ad). When time is pressing, these preparations render useful service, and producing the colour on the metal itself and not on a "foreign" coating, are infinitely superior, as blueing solutions, to the copper sulphide method described under "Retouching" and useful only on emergency for the latter purpose. Whether the coating given by "express" browning is as dense and durable as that produced by the orthodox methods, can only be determined by prolonged experience, and must be considered as yet "not proven."

Troubles. It is at this stage that troubles are usually encountered, and should be gotten rid of before proceeding further. The commonest is refusal to "take" the solution in places, then irregular or difficult rusting, and finally an unsatisfactory appearance when finished.

Refusal or irregularity in "taking" the browne is usually due to incomplete degreasing, which has already been dealt with: difficult rusting is traceable to wrong temperature conditions, the physical state of the surfaces, or unsuitable composition of browne in relation to that of the metal. More liberal application of the browne, or reapplication after the first application has dried, has also been alluded to.

The influence of temperature is clearly shown by the fact that many brownes act with extreme slowness at ordinary temperatures, but in every way satisfactorily at that of the steam oven (examples in Chapter IV): in other cases preliminary warming of the parts to 60-70°C. (140-160°F.) before applying the solution, instead of applying it cold, will prove beneficial, but no fixed rule can be given.

There is no "universal" browning solution: its chemical composition must be governed by that of the metal. This is specially the case with the newer alloy steels, according to the special constituents giving them their value. Crude or fuming nitric acid or aqua regia instead of pure nitric acid will help with difficulty oxidisable steel: high-chromium steel resists nitric acid, but can be browned (although unsatisfactorily) with solutions containing hydrochloric acid.

The physical condition of the metal surfaces is of influence, and in general, highly-polished parts take the solution less easily than dull ones: the use of slight pickling, in order to facilitate initial action, has already been mentioned under "Degreasing."

Slight preliminary pickling is often a cure for an unsatisfactory final finish: in certain cases it will assist in producing a fine deep blue- or ebony-black tone where otherwise a displeasing chocolate- or reddish-brown has resulted, so that in such cases this "kink" should not be left untried.

Reconditioning.   Already blued or browned parts that have become rubbed bright in places by wear can be rebrowned without repolishing the entire surface, and with a little care no irregularity in the colouring can be detected: this is best assured by the use of a browne giving a deep black tone. Uniformity is naturally more easily assured by repolishing the entire surface.

Nickelled parts.   These must be denickelled before browning, as described in Chapter VI.

Inlaid parts.   The barrels of luxury arms frequently bear the maker's name in gold or silver inlay, other parts, e. g. the frames of double guns or express rifles being decorated in similar manner. Such pieces are browned in the usual way, but the inlay requires certain reservations.

In the first place, brownes containing *any* salt of mercury deposit a thin film of this metal on gold or silver, with which it eagerly amalgamates (not with iron or nickel) : on gold the slight silvery film is easily detected and when rubbed shows up bright. It can be driven off by heat, but this is almost certain to loosen the inlay, or by repeated touching with a pointed stick dipped as necessary in nitric acid: an extremely delicate operation, as it is almost impossible to avoid contact with the already-browned, surrounding iron, damaging the browning. Prevention is here emphatically preferable to the best cure, and all mercurial brownes banned accordingly.

Nitric acid brownes attack silver, so that only hydrochloric

ones must be used with silver inlays, and brownes containing
aqua regia are to be omitted.

Penetration of the browne between metal and inlay should
be excluded by quality of workmanship, this being de luxe
work, but if it occurs, any harmful effect is usually destroyed by
the subsequent boilings (slightly alkalising the water if necessary): here again brownes inclining to after-rusting are best
avoided.

Scratching in the neighborhood of the inlay should be done
with special care and a soft iron wire brush, to avoid deep
scoring of the soft metal: inlaid pieces thus require special care,
corresponding to their luxury-character.

Malleable iron. This is usually less easy to brown satisfactorily as the material differs from forged parts in being distinctly porous: the finish therefore tends to "sweat," resulting
in a patching or even a pitted surface.

Hardened and casehardened parts. These are in general
less easy to brown than unhardened pieces, due to physical
difference: hardened or tempered parts however, being uniform,
end by taking as good a finish as unhardened pieces.   Casehardened parts take the browne variably, and it is often difficult
to initiate action.   Even when uniformly polished before browning, a variegated appearance of the finished article, resembling
faintly that of marble, is almost always perceptible, especially in
glancing light and on relatively large surfaces, such as Martini
and Winchester type receivers.

This, in any case slight and not very noticeable "beauty fault"
can moreover be eliminated with ease and certainty, by "priming" the pieces with weak (1 to 2, exceptionally 5%) salammoniac solution, applying the browning solution proper only
when rusting has started.   "Priming" is only needed on first
application, rarely on repeating: it can be used more generally to
advantage to promote rusting, e. g. when using very slow-acting
brownes, in dry, cold weather, etc.   (See moreover Chapter 4
"Salammoniac browne, Aa2.)

This treatment enables a fine, uniform and deep black brown
to be obtained on casehardened as easily as on nature-hard parts.

5. Darkening

The effect of boiling or steaming the rusted parts has already
been described, also the greatly facilitated work of scratching,
and the improved appearance of the finish.

For the individual gun owner and general gunsmith boiling

is preferable, no special equipment beyond a tank of sufficient size and the means of heating being required. Steaming is the rule in large-scale production, the final effect being identical in either case. The duration of boiling or steaming is generally from 15 to 20 minutes, but the conversion into magnetic oxide is mostly if not entirely accomplished in the first 5 minutes.

Water. Whenever possible, soft water only should be used for boiling, as the carbonates in hard water tend to reconvert the magnetic oxide, thereby producing a brownish to fox-reddish colour, in general proportion to the hardness and with certain solutions more than with others. Where distilled water is not specially cheap and abundantly available, filtered rain water is best, unless acidified by sulphurous furnace gases or the fumes of near-by chemical works. Engine condensate, always contaminated by oil, must not be used, but steam turbine condensate (except that from exhaust steam turbines coupled to primary engines) is free from oil.

Bores. The sealing of barrels by wooden plugs has already been mentioned: note however that these swell on boiling or steaming and contract after cooling, and that temperature changes and expansion of the air in the bore help in loosening them, so that with a little carelessness water almost always ends by finding entrance into the bore. Air pressure in the bore can be prevented by inserting into one of the plugs a narrow tube bent up above water level, but this is rather troublesome and not a complete cure on account of the other factors mentioned.

The penetration of plain boiling water into the bore is however not very harmful, as a very small quantity of *caustic* potash or soda suffices to neutralize all acid residue taken up by the water, thus avoiding all harm to the bore. (As detailed below, this addition is sometimes desirable on other grounds.) For a considerable period the author's regular practice has been to make this addition when boiling and to leave the bore open.

If however plugged in the usual way while alkalising as described the plugs should be renewed, or at least tightly driven in at each processing.

Acid baths. Conditions are different when the barrels are boiled in an acid bath (e. g. in certain methods of treating stainless steel) and the entry of liquid into the bore must be positively prevented. In this case recourse must be had to tightly driven plugs of soft metal, providing one of them with an equalising tube. (See under "Special cases. Rustless steel.")

Operation. Boiling needs no special description, and steaming in an oven at higher temperature offers no further difficulty where installed. Barrels and other parts of inconvenient shape for the private gun owner can easily be steamed without special equipment, by slowly moving them to and fro before the spout of a common kitchen kettle, turning them round continually so that the entire surface is subjected uniformly to the action of the steam. It is important to keep the kettle boiling *briskly,* otherwise the rust, instead of being converted and loosened, is rendered if possible even more adherent, partly discoloured, and very difficult to scratch.

But given the necessary care—*and* patience—as efficient work can be done with this kitchen makeshift as in the best-equipped Armoury or custom gunmaker's workroom.

Variations. The usual practice is to boil (or steam) and scratch after each pass (successive coating), but certain recipes prescribe boiling after alternate passes, or only once after the last: the common practice however appears to be the most logical, and there is little to be said for alternate boiling and scratching unless the coating produced by a single application is exceptionally faint.

Occasionally, and in particular with twist barrels, a little ferrous sulphate and logwood (Fr. "bois de Campèche," G. "Blauholz") chips (abt. 30 grs. per U. S. quart) are added to the last boiling, giving a slightly reddish-blue tint to the coating, but this is quite exceptional for steel barrels.

6. Scratching

(Fr. "cardage," "grattebossage," G. "Kratzen.") The loose rust formed is now removed by means of a wire brush, or according to the time-honoured term, "scratched": at successive rustings the coating takes an ever deepening tone, passing from grey to brown and black: usually from 2 to 4 repetitions are sufficient. As already mentioned, the fox-red rust in brown browning in far harder to remove than the boiled or steamed magnetic oxide, and as the beauty of the final finish depends greatly on the thoroughness of scratching, the argument for scratching at each pass needs no further emphasis.

Scratching is done by hand brushes of iron or steel wire, of size and shape best adapted to the particular gun parts, in trade gunmaking and in Armouries by the much more rapid circular brushes. Where power is not available for driving, long pieces such as barrels are far more easily scratched with steel wool than with the hand brush, and for this purpose the very fine quality used by cabinetmakers should be chosen. Brush and steel wool give exactly the same excellence of finish, the ever-necessary patience pre-supposed.

Scratching is best effected on exactly the same lines as filing, using first a coarser, then a finer brush, and scratching lengthwise and crosswise alternately. The final scratching of barrels is therefore best done crosswise, as the hair-ridges which occasionally remain distinguishable (but *should not,*) more effectually prevent annoying light-reflection.

Hand brushes are mostly of crinkled steel wire, circular ones of straight or crinkled, soft or spring-tempered steel, occasionally also of iron wire. A thickness of wire of about 6/1000ths" is the usual average, but one exceeding 8/1000ths" is unsuitable, as with a very little carelessness it simply wipes away all the colour already formed, especially during the first coatings. Brushes of finer wire, from 2 to 3/1000ths", also the gentler-acting ones of iron wire, are recommendable for the final coats, to the advantage of the finish.

The rule for operating circular brushes is identical with that for emery wheels: light touch and a peripheral speed of 25 to 40 ft./sec. ensure quick and effective work without undue wear of brushes. Emery-wheel speed (average 83 ft./sec.) is not only unnecessary, but on account of its over-keen action easily results in patchy work, and causes excessive wear of brushes. An average peripheral speed of about 33 ft. sec. with a

brush diameter of	6	8	10	12 inches
corresponds to about	1260	945	755	630 revs./min.

Circular brushes with wooden centres should only be mounted on cylindrical spindles, as if carelessly driven on taper ones the centre is easily split. Brushes with metal centres are unconditionally preferable. For continuous work brushes with replaceable wire armature are strongly to be recommended: these consist of a number of flat discs with wire brush insets, pressed side to side between flanges, or of a centre into the circumference of which a series of small flat brushes is dovetailed, after the manner of certain types of steam turbine blading.

The shape of certain pieces compels scratching with the hand brush: otherwise the circular brush is greatly preferable on account of its immensely speedier action. Its only disadvantage is the production of a whirl of almost impalpable dust, injurious to lungs and soiling the surroundings: it is therefore best adapted to power-driven shops where the provision of dust-collecting devices offers no difficulty and is in fact a self-understood necessity.   Smaller concerns and private users who only brown occasionally can evade the dust plague by scratching out of doors, the inconvenience of which needs no special mention.

Keeping circular brushes in order.   The wires gradually wear away by use, thereby becoming stiffer and relatively harder, and when much worn can easily wipe away the coating already formed.   They also first bend, then get interlaced and entangled, finally into hard knots by continued neglect, until the brush becomes completely unusable.   To prolong their useful life as far as possible, the brushes should be reversed occasionally on the spindle while the wires are still only slightly bent: if this bending has gone further, they should be straightened by combing out with a stiff metal comb, currycomb or even a sharp nutmeg grater.   If however knotting has gone too far for this remedy, the only cure is lopping with shears, after which the brush is indeed of little further use.

That all scratch brushes, and the steel wool otherwise used, must be strictly kept from all soiling by oil or grease, need not be further insisted on.

F. Brandeis ("Die moderne Gewehrfabrikation") very justly emphasizes that the beauty of the final finish depends in large measure on the thoroughness of the scratching, recognisable by the smooth grey to deep black surface free from any speck of dust—even when viewed with a magnifier.   Therefore patience and again patience.   Observing the motto "thorough" at any stage enables the next step to be undertaken with a clear conscience.

Careless or incomplete scratching, on the other hand, betrays itself by uneven, more or less patchy appearance, and any neglect at any stage is ever so much harder to make up for subsequently, than the extra trouble needed for prevention in the earlier step.

7.   Neutralising

This is only needful now and again, as certain browning solutions, among them some of the very best, have the unpleasant fault of causing after-rusting of the finished parts.   The cure is however of extreme simplicity, and consists in the addition of a small quantity (30 to 45 grs. per U. S. quart) of *caustic* potash or soda to the last boiling water.   If the parts are steamed instead of boiled, soaking immediately after scratching in weak (4 to 5%) caustic alkali lye, either for a few hours in gentle warmth (30 to 35° C. $=$ 85 to 95° F.), or overnight in cold lye.   In the absence of *caustic* alkali, neutralisation can be accomplished by soaking over night in 20 to 30% potash or soda (potassium or sodium *carbonate*) solution, but this must only be used cold: if heated or, worse still, boiled, the carbonates (almost always present to a very slight extent in caustic alkali) partly or entirely convert the carefully-formed black oxide into fox-red carbonate: a most unwelcome accident that is hard to repair even with a coarse scratch brush, when it does not mean the entire ruin of the mishandled pieces. Moral—for the chemically un-initiated—look out for mistaking chemical names. Rightly applied, the neutralising method here given does no harm whatever to the browning.

Although this after-rusting tendency is in itself undesired and undesirable, it is not without a certain utility—not altogether irreproachable—as a means of obaining a dull finish. After-rusting goes on under the oil applied to "fix" the coating, although more slowly on account of the exclusion of atmospheric oxygen: if therefore the finished but *not* neutralised parts are allowed to stand, they after-rust and at a point only recognisable by inspection and experience, are again degreased, scratched, neutralised and fixed, they then exhibit a clean, dull surface.

The colour tone remains unaltered, but agreeably dull on account of the slight etching action. Frequent inspection and experience are needed to recognise the point at which to arrest further action, which would cause pitting, but rightly carried out, this method gives excellent results.

8. Retouching

It occasionally happens that through inadvertence or other causes certain parts of the surfaces do not "take" the browne and remain bright on finishing. Such small defects can be remedied in various ways, but the only really satisfactory method is to apply a quick-acting browne and carry on in the usual way. Only a portion of the surface adjoining the unbrowned spot need be degreased: darkening is done by local steaming, repeating as requisite. When carefully done the repair cannot be detected, and having been effected on the metal itself, the finish is as durable as that of the remaining, originally browned surface.

A quick but less recommendable makeshift is the following. The bright spot—duly degreased—is moistened with saturated (abt. 2 g. in 10 ccm. = abt. 92 grs. in 1 U. S. fluidounce) copper sulphate solution, a deposit of metallic copper immediately forming. Alternate drying and re-moistening is repeated a few times, then when dry, the coppered spot is painted with ammonium sulphide (or "liver of sulphur") solution, forming in a few seconds black copper sulphide: let dry and repeat a few times, then polish with woolen rag and oil.

Another method requires the use of the above solutions and in addition, one of lead acetate ("sugar of lead") and of sodium hyposulphite (the photographer's "hypo"): more complication and result no better. These processes are advertised under various fancy names (and prices) as quick browning solutions, but while serviceable as temporary makeshifts, they are in every way inferior to regular browning, because not producing the colour on the basic metal: moreover, the copper adheres but slightly, is liable to peel, and its sulphide is far less wear-resisting than the oxide of regular browning.

9. Fixing (Finishing)

After the final scratching (and neutralising when necessary), the browning is rendered permanent by simple oiling, which inhibits further action on the metal. Generally the colour then appears to become slightly lighter, although this may simply be an illusion due to the difference between a dull and a bright surface.

With after-rusting solutions, the author's practice has been to oil the parts with vaseline oil (L. "paraffinum liquidum," b. p. 360° C. = abt. 680° F.) to a temperature of 130 to 150° C. (abt. 265-300° F.) as a maximum, until the oil has evaporated, when the deep black tone, perhaps even slightly deepened, remains. Then rub off any oil residue and oil for good.

This incidentally provides a check on the thoroughness—or the reverse—of the scratching. If thoroughly done, no loose particles of rust remain on the surface after the oil has evaporated, and conversely. The temperatures stated are far below that of ordinary annealing and are only maintained for a short time, so that this treatment is quite harmless to springs and other tempered parts. A further useful effect is to drive off with certainty any adhering residue of acid or damp: the author has thence adopted this very simple finishing operation as standard practice, irrespective of the particular solution used.

10. After-treatment

Black-browned (blued) parts are now finished and need no further attention. In many cases brown-coloured parts require further treatment, either to improve appearance, or to prevent after-deterioration: examples of the first are twist barrels, to bring out the pattern, and those browned with antimonious solutions, which tend to after-rust, on account of the very differing electric potential of iron and antimony.

Waxing. Parts of simple shape are rubbed handwarm with pure, yellow or bleached beeswax (or wax substitute, e. g. ceresine), then successively with a bristle brush, wash-leather or woollen cloth, and polishing stick or burnisher, until all wax excess has disappeared and the pieces feel quite dry to the hand after holding them for a few seconds, if still feeling greasy, rubbing must be continued, otherwise the surface will remain finger-marked.

Parts of complicated shape are best coated by means of a spray (also convenient for waxing brush or rag) and correspondingly treated. The following are a few well-tried solutions, of which many other compositions are to be found in technical manuals.

a.	Beeswax	6	parts by weight
	Benzine	94	" " "
b.	Curd soap	4	" " "
	Japan wax	12	" " "
	Water	84	" " "
c.	Beeswax	33	" " "
	Turpentine (hot)	67	" " "

Solution (a) is for spraying: extremely inflammable, so keep away from flame: (b) is prepared by boiling together and stirring till thoroughly mixed and cold.

Waxing imparts a certain translucence and a mild, inconspicuous sheen to dull or earthy-looking surfaces, fills up the pores when thoroughly done and excluding penetration by atmospheric oxygen and damp, improves the rust-protecting value of the coating. Oil can also be used instead of wax, but with duller and less permanent effect.

Lacquering. This is now comparatively seldom done, except on casehardened or other parts of somewhat earthy appearance. Cellulose lacquers (G. "Zaponlack"), a solution of celluloid or cellulose nitrates in ether-alcohol, acetone, amyl acetate, etc.) and its variants, are particularly suited to this purpose: they adhere strongly, are colourless and inconspicuous, therefore do not alter the aspect of the base metal and resist wear well.

Cellulose lacquer is applied to the pieces by brush, dipping or spraying—special trade varieties are available for these modes of application—and dried at room temperature or in ovens at 35 to 50° C. (abt. 95 to 120° F.). Greasy surfaces or over-diluted solutions give rise to rainbow colours, and lumps may form on dipped parts or by excess application, although this can be easily avoided by letting them drip.

Cellulose lacquer is unsuitable for pieces coloured by anti-monious or arsenical preparations, as these are then rendered patchy: spirit lacquers are alone suited to such.

The following are a few older well-tried lacquering composi-tions for barrels ("Stelle & Harrison"):

a.	Shellac	137	grs.	3.0	g
	Dragon's blood	23-34		0.5-0.75	
	Alcohol 95° to	10	fl. oz.	100	ccm

This is for antimony-browned barrels. Dragon's blood according to shade desired.

b.	Shellac	27	grs.	0.6	g
	Sandarac	27	"	0.6	"
	Venetian turpentine	5	"	0.1	"
	Alcohol 98° t. s. to	10	fl. oz.	100	ccm
c.	Clear mastic grains	457	grs.	10.0	g
	Camphor	228	"	5.0	"
	Sandarac grains	45	pieces	15	pieces
	Elemi	228	grs.	5.0	"

Alcohol 95° t. s. to dissolve.
Used cold. Elemi toughens.

Spirit lacquers demand noticeably greater skill in using if appearance is to be satisfactory: too-sparing application causes rainbow colours in streaks, too-liberal causes unsightly ridges: the parts must mostly be warmed to about 60° C. (140° F.), otherwise the coating remains dull and opaque.

As in browning, thorough degreasing is an essential pre-condition of satisfactory lacquering.

11. Special cases

a. **Twist Barrels.** With the formerly universal, but now ever decreasingly used twist barrels, it remained to bring out the beautiful wavy pattern of the twist, an effect only faintly marked by the process of browning: the harder steel strip is less acted on than the iron, so that it stands out slightly, producing alter-nations of lighter and darker waves.

This effect can be further heightened by the pre-treatment described in Chapter V.

Generally speaking, weaker solutions are alone adapted for browning twist barrels, on account of the want of uniformity of the material. Slow action and many rustings are the correct procedure here, the more as brown finish was the rule, and the brown oxide is in any case much harder to scratch. Barrels are therefore neither boiled nor steamed, but only scalded once when the desired tone has been reached after the last rusting: occasionally they are laid for a short time in hot *(not* boiling) decoction of logwood (about 30 grs. per U. S. quart) sometimes adding a little copper sulphate. The effect is to produce a warm, slightly blueish, plum-brown finish.

After scalding, the barrels have a uniform colour, on which the pattern shows more or less distinctly according to the effect of the solution, but it usually requires additional treatment to bring out its full beauty.

This can be effected by rubbing with a polishing stick, woolen or linen rags and chalk paste, very fine pumice stone or flour emery and water (the last very keen: caution!), or by careful rubbing with Arkansas or other very fine oilstone. Then wipe clear, wax, oil or lacquer. As may be readily understood, a high degree of skill and experience is essential to the production of a regular, beautiful finish.

Chemical methods attain a similar result much more easily, as described in Chapter V, under "Etching twist barrels."

The above summarises the general course of finishing twist barrels, but almost every gunsmith has his own particular way, the result of long experience.

Note. Solutions described in Chapter IV as intended for "twist barrels," must be understood as having been specially applied to them in former days for the production of brown finish, but (like all others) give a lighter or deeper black after boiling or steaming the pieces.

b. "Stainless" steel barrels. To colour "stainless" or "rustless" steel by rusting is apparently a gross contradiction, more apparent however than real. The name "rustless" has here about the same value as the designation "smokeless" applied to nitro-powder: "slight-" or "hard-rusting" (the equivalent of the very correct German names) would be truer to fact if less desirable for selling.

So-called "Stainless" or "Staybrite" steel (in the main a high-chromium steel, also with various additional components) is in fact a metal of very small oxidisability, but not absolutely un-oxidisable: it can be rusted by many solutions, for example. The author has for instance obtained a very good (but evanescent) deep black on stainless steel with the Creusot browne (Chapter IV, Group Ac. 1.), although very slowly and after many rustings. Nitric acid solutions are useless, as it is without action on this steel, while violently attacking almost all other metals. Two further methods are described in Chapter IV,G. but like the preceding, are accompanied by faults that greatly detract from their value.

The trouble with regular browning is that the rust adheres very slightly, and even gentle scratching with soft iron wire brushes easily and sometimes completely wipes it away, thence slow, uncommercial working. Dipping methods have the grave fault that absolutely certain protection of the bore is a serious difficulty: paraffin-soaked plugs will suffice for the chromic acid bath, but the heat of the permanganate one requires plugs of soft metal. Moreover, both methods leave much to be desired as regards appearance of the finish.

A further defect of these processes is the instability of the coating, which exhibits the peculiar behaviour of slowly fading when vaselined as usual for storing. The author had certain stainless barrels browned by the permanganate method, and where wrapped in vaseline rags, as over fore- and rearsight bases, the black colour gradually faded and had practically vanished after 4 years, and this effect seems to extend, only more slowly, over the entire surface of the barrels. On two barrels the colour was moreover patchy, somewhat recalling the markings of a tabby cat, but this may of course have been simply part of the "teething troubles" of the process: others were a good deep black. Permanence thus leaves much to be desired, and so far as the author's experience has gone, no solution has given better results.

The general conclusion to be drawn from these—comparatively few—trials is that the regular browning method is unsuited to the material, which is of special value to the private shooter, and for machine gun barrels. At the present time the following methods are in use for overcoming the difficulty:

    a. Temper-blue, that is, simple oxidation by heat. The simplest method. As in all similar cases, the layer of oxide is of sub-microscopic thickness, or rather thinness, and even less wear-resisting than that formed on ordinary steels: conclusion, not recommendable.

    b. Coppering, preferably by electroplating, then colouring the copper deposit. More durable than (a), but still not recommendable, because not formed on the barrel metal itself.

    c. Black nickelling, also electrolytically. The nickel coating is black throughout, adheres strongly, is very durable on account of the hardness of the metal, and of very pleasing appearance. Far superior to the two preceding.

    d. Iron-plating, (G. "Verstählen"), also electrolytically, then regular browning of the deposited iron coating. The most rational theoretically, and when carefully carried out also the best practically in respect of durability and appearance.

Note. According to Dr. Pfanhauser ("Galvanotechnik," Berlin 1928), durable electroplating of stainless steel can only be attained after superficial dechroming of the metal: this can be accomplished electrolytically, the piece forming the anode, or more simply, by laying in warm (abt. 50° C., say 120° F.) 30% hydrochloric acid, until distinct evolution of gas sets in.

Treatments (c) and (d) are naturally best carried out by special electroplating concerns having scientific staff and the necessary special equipment.

12. Protective Value of Browning

Common experience proves that browned gun parts rust less easily than polished ones, although the protective value of the coating is neither unconditional nor complete. In the first place it is porous and therefore penetrable to atmospheric damp and gases: secondly, metals being electropositive to their compounds, a galvanic couple, to the disadvantage of the metal, is always present. If the pores of the coating are well filled by rubbing in wax, paraffin, vaseline or mineral oil, atmospheric oxygen, damp and other destructive agents cannot reach the metal, so that direct rusting is prevented and the galvanic couple is deprived of the opportunity of exercising its injurious influence. Close grain and a strongly adherent coating, obtainable by suitable choice of browning solution and procedure, are therefore always to be aimed at.

Following this, periodical, thorough waxing, vaselining, etc., of the browned parts greatly assists in keeping the arms in perfect condition. Note however that mere dry dust, if uncontaminated by chemical impurities, such as collects in Armoury racks, is harmless.

Much more efficient rust-protection is afforded by the phosphatising process described in Chapter VII, but even this is not absolute and is only occasionally used for gun parts.

13. Removing old Browning

This is often necessary in reconditioning: mechanical removal, where no tumbling barrel or bowl is available, is only applicable to small parts of simple form, otherwise is slow and troublesome, and to be avoided. Preliminary cleaning of old coating can be done with a circular brush of coarse wire (10 to 12/1000ths thick), but chemical cleaning is both easiest and complete: this is more fully described in Chapters V and VI, "Etching and Derusting."

CHAPTER III

PREPARATION OF BLUEING AND BROWNING SOLUTIONS

1. Elementary Chemical Explanations

Browning is essentially a chemical operation and is based on the action of acids and salts on the metal in presence of atmospheric oxygen and moisture. No direct chemical knowledge is necessary for the preparation of blueing or browning solutions, as the gunsmith and private owner can always get them made to formula by the nearest druggist, and whoever desires to make his own has only to follow the directions and data given hereafter. For the benefit of those who desire to know more of the "why and wherefore," but to whom chemical notions are unfamiliar, the following simple explanations are offered, which the chemically-initiated can quietly pass over unread. The particulars and instructions given in the Special Division will enable anyone to make up his own preparations with insignificant trouble.

Elements and Compounds. (*). Chemically considered, the universe consists of two kinds of bodies, namely, simple and composite, in chemical parlance, elements and compounds. Elements are bodies consisting of a single kind of matter, incapable of being split into different components by any means whatever. Compounds can on the other hand be resolved into their constituent elements by appropriate means.

Up to the present about 92 elements in the universe have been recognised or isolated, some in so to speak immeasurable mass, others in the tiniest traces. They fall into two categories, metals and non-metals, whose nature (in the popular sense)

(*). Modern research has greatly modified the older chemical theories, which however completely suffice for the present purpose.

needs no explanation.   The elements are nearly all solids: at the ordinary temperature only two (bromine and mercury) are liquid, and of the commoner ones five (hydrogen, oxygen, nitrogen, fluorine and chlorine) are gases.   Only those useful for the purpose of this work will be here mentioned, also their Latin names, where very different from the English, and the symbols by which they are designated for the sake of brevity in chemical work.

Name	*Symbol*
Hydrogen.	H
Oxygen.	O
Nitrogen.	N
Chlorine.	Cl
Antimony (Stibium).	Sb
Bismuth.	Bi
Carbon.	C
Chromium.	Cr
Copper (Cuprum).	Cu
Lead (Plumbum).	Pb
Manganese.	Mn
Mercury (Hydrargyrum).	Hg
Phosphorus.	P
Potassium (Kalium).	K
Silver (Argentum).	Ag
Sodium (Natrium).	Na
Sulphur.	S
Tin (Stannum).	Sn
Zinc.	Zn

To these must be added the radicals "Ammonium," symbol NH_4, (popular abbreviation "Am"), which behaves as a metal, and "Cyanogen," symbol CN (popular abbreviation "Cy"), which acts as a non-metal.

The first 4 of the above list are gases, their names, derived from the Greek, meaning respectively "generator of water, of acid, of nitre," and "green," from the greenish-yellow colour of chlorine.

Chemical action.   Chemical actions—termed "reactions"— take place in virtue of "affinity" (attraction), which is differently manifested in each particular case, but is strongest between metals and non-metals.   As a universal principle, chemical action is based on the "right of might," a given body having stronger affinity under given conditions expelling one of lower affinity from its compounds.   An iron rod dipped in solution of copper sulphate, for example, becomes covered in a few seconds with a film of metallic copper, the affinity of iron to the acid radical being stronger than that of copper.

Primary chemical compounds. Non-metals combine both with metals and with each other. Combinations of oxygen (partly also of sulphur) and metals are generally called "bases," those with non-metals are designated "anhydrides," which in turn combine with water to form "hydrates" and "acids."

Hydrogen (which behaves in many cases like a metal) also combines with certain non-metals to form compounds with properties varying between strongly basic to equally strongly acid character. Acid compounds of this description are also known as "hydracids," in distinction from "oxyacids," which also contain oxygen.

Secondary chemical compounds. Bases and acids having opposite chemical character, combine with each other, each losing its characteristic qualities, or in chemical language, "neutralising" each other. Such compounds are called "salts," after their best-known prototype, common salt.

Bases and acids are empirically distinguished from each other in that bases turn red litmus paper *blue*, while acids turn blue litmus *red*.

Salts are in general neutral, but "basic" and "acid" salts also exist, in which the acid or basic radical is only partly neutralised by its counterpart, so that a basic or acid residue remains in the salt: examples are basic bismuth chloride, acid sodium carbonate.

Neutral salts do not change the colour of litmus, unless a strong base is combined with a weak acid, or vice versa. Thus, neutral sodium carbonate (common washing soda) reacts strongly basic, ferric chloride strongly acid, but acid sodium carbonate nearly neutral.

Insoluble bases and acids do not affect litmus paper: iron oxide, which combines with acids, is nevertheless a base, silicon oxide (pure sand, quartz) which is quite neutral at ordinary temperature, is the strongest of all anhydrides at white heat, then expelling all other acids from their salts.

Acids and salts are the chief substances used in browning.

Oxidation and reduction. "Oxidation" in its strict sense means the union of any body with oxygen. Examples, all forms of combustion, fermentation, the rusting of iron, etc. The opposite action is called "reduction" (deoxidation), e. g. the smelting of metal from ore, in which the higher affinity of the glowing fuel for the oxygen of the ore liberates the metal.

Under certain conditions various metals form different oxides with progressively increasing oxygen content: the passing of a lower to a higher oxygen content is also an oxidation, and quite generally if not strictly accurately, the conversion from a lower to a higher non-metallic content—other than that of oxygen—is colloquially spoken of as an "oxidation," e. g. the conversion from *ferrous* to ferr*ic* chloride: and conversely, the opposite process is termed a "reduction."

Chemical names. Positively *accurate* knowledge of the meaning of chemical names is of paramount importance for this and every other purpose, the more as names differing by only a syllable or even a single letter may mean bodies of the most divergent chemical nature.

Chemical nomenclature has gradually grown with the progress of the science, so that at the present time many ancient, often unfortunate and even positively misleading names are still in trade use besides the scientific ones. Here only the most careful attention on first acquaintance, then acquired habit, and memory, can help. The following indications can be taken as a general rule.

Bases. Compounds of metals with oxygen, not containing combined water, are called "oxides," those containing such (as often happens on precipitation) "hydroxides" or "hydrates."

Compounds of metals with chlorine are called "chlorides," with bromine, "bromides," with sulphur, "sulphides" etc.

Where a metal forms several oxides, that with the lower oxygen content is distinguished by the name of the metal with the suffix "-ous," and with the higher content by the suffix "-ic": thus mercurous and mercuric oxides, ferrous and ferric oxides. Where the common name of the metal does not lend itself to this distinction, the Latin name is substituted, as in the example above. The older terms "protoxide" (proto-oxide, meaning first oxide) and "peroxide," are still in common use in English and French, but the latter term often means a still higher oxide, the syllable "per-" being a contraction of "hyper-," meaning "beyond" the normal high limit. Manganese, for instance, forms 3 basic oxides, properly speaking, namely, manganous, mangano-manganic and manganic oxides, then a peroxide (and finally an "ultra" compound (permanganic anhydride).

In French, the lower oxide is distinguished by the adjective composed of the name of the metal with the suffix "-eux," the higher by the adjective ending in "-ique:" thus "oxyde cuivreux" and "cuivrique," similarly for chlorides, sulphides, etc.

In German, the lower oxide of a metal is called "Oxydul," the higher "Oxyd," and similarly "Chlorür" and "Chlorid," "Sulfür" and "Sulfid:" these explanations will serve to prevent misconceptions in the literature of these languages.

Acids. Certain non-metals also combine with oxygen in more than one proportion, forming as many different anhydrides and acids.

The principal (so to speak "normal") acid of a non-metal is called by its name with the suffix "-ic," e. g. sulphuric, nitric, phosphoric acids, etc. The next lower acid is distinguished by the suffix "-ous," e. g. nitrous, phosphorous, etc. acid, and a still lower one by the prefix "hypo-" (meaning "under") and the suffix "-ous," e. g. hypochlorous, hypophosphorous, etc. acid. A higher than the "normal" acid is similarly distinguished by the prefix "per-," as for example, perchloric, persulphuric, permanganic, etc. acids.

In French and German these differences are expressed as shown by the following equivalents of the English:

Nitric acid.	Acide nitrique.	Salpetersäure.
Nitrous acid.	do. nitreux.	Salpetrige Säure.
Hypochlorous acid.	do. hypochloreux.	Unterchlorige Säure.
Permanganic acid.	do. permanganique.	Uebermangansäure.

Salts. Neutral salts of a metal "M" with the radical of a non-metal "X" generally follow the rule here given:

Salt of a "normal" acid	M-X-ate.
do. 1st subnormal acid	M-X-ite.
do. 2nd subnormal acid	M-hypo-X-ite.
do. an over-normal acid	M-per-X-ate.

examples, zinc sulphate, potassium nitrite, sodium hyposulphite, potassium perchlorate.

Basic and acid salts. The common names of these do not follow any settled rule, consequently are very confusing to the non-chemist, and only habit and memory can help. Acid sodium carbonate is called "bicarbonate," but potassium "bichromate" is not acid potassium chromate, but the normal (neutral) salt of *di*chromic acid, a different body. "Sodium phosphate" is not the neutral salt, but the 1/3-acid salt of phosphoric acid, and basic bismuth nitrate is known as bismuth "subnitrate".

Similar uncertainty exists in regard to the prefix "per-." Manganese peroxide is really a "per-oxide" of manganese, but "iron peroxide" is merely an archaic term for "ferric" oxide, and "ammonium persulphate" is the salt of *per*sulphuric acid, a product of modern times.

Latin names. Many chemical dealers and works issue their catalogues with these names, which are used in all pharmacopoeias and have the advantage of meaning the same thing (with the rarest exceptions) in all countries—their use when buying or ordering chemicals is therefore strongly recommended as a precaution against mistakes. Abbreviations however present pitfalls, and the following typical examples of names should be carefully memorised until their different sounds and meanings have become thoroughly familiar.

Thus:

Potassium chlor*ide*	is called	Kalium chlor*atum.*
" chlor*ate*	"	" chlor*icum.*
Sodium nitr*ate*	"	Natrium nitr*icum.*
" nitr*ite*	"	" nitr*osum.*
" *hypo*chlor*ite*	"	" *hypo*chlor*osum.*
Ferr*ous* chloride	"	Ferrum chlor*atum* oxy*du*latum.
Ferr*ic* chloride	"	" oxy*datum.*
Potassium sulph*ide*	"	Kalium sulfur*atum.*
" sulph*ate*	"	" sulfur*icum.*
Sodium sulph*ite*	"	Naturium sulfur*osum.*
Ferr*ous* sulphate	"	Ferrum sulfuricum oxy*du*latum.
Ferr*ic* nitrate	"	" nitricum oxy*d*atum.
Potassium *per*chlor*ate*	"	Kalium *per*chlor*icum.*
" *per*mangan*ate*	"	" *per*mangan*icum.*
" cyan*ide*	"	" cyan*atum.*
" cyan*ate*	"	" cyan*icum.*
Acid potassium sulphate	"	Kalium *bi*sulfuricum.
Acid sodium carbonate	"	Natrium *bi*carbonicum.
Antimony oxy*chl*oride	"	Stibium oxy*chl*oratum.
Basic bismuth nitrate		Bismutum *sub*nitricum.

These names are frequently abbreviated by the omission of the final "-um," thereby sometimes closely resembling the common names of other bodies having totally different chemical character. Such must be specially warned against, as very unpleasant results may follow misunderstandings.

If for instance "kali. chlorat." is misunderstood to mean that potassium chlor*ate* is wanted and accordingly supplied, violent explosions may conceivably result, while if instead of the relatively innocuous "potass. *cyanate*" "kali. cyanat." is ordered, a most violent poison is received, to the common danger. This by way of illustration, neither substance being used for browning.

Recapitulation. This chapter can be summarised as follows:

1. Bases and acids are primary compounds of opposite nature which combine to form salts, each component neutralising the other's characteristic properties.

2. Salts are in general neutral substances, but "basic" and "acid" salts also exist, in which the base or acid is only partially neutralised.

3. All blueing or browning mixtures consist chiefly of solutions of acids and salts.

4. Their action is based on the higher affinity of iron for their acid content and the radicals of the contained salts.

5. Oxidation and reduction—popularly but not strictly correctly to be compared to rusting and derusting—are opposite actions of universal application in everyday life.

6. The greatest attention must be paid to the *exact meaning* of all chemical names.

2. Workroom Quantities

For the private gun owner and gunsmith who only browns occasionally, a quantity of 4 fl. oz. of a few selected solutions will suffice for months: quart—or gallon—wise preparation is only needed for more or less wholesale operation. Certain solutions are not stable and will not keep indefinitely, and are thus best made up in small quantities: the lower limit can be put as about 2 fl. oz., otherwise weighing out the very small quantities of chemicals becomes troublesome.

Care exercised in making up pays for itself here as everywhere, but over particular accuracy, e. g. using an analytical balance, is unnecessary, first, because the chemicals used are not absolutely pure, secondly, because the process is a technical one, and minute differences in composition are without effect on such.

3. Equipment

Very little is needed in this line: a sufficiently accurate but not over-sensitive balance (e. g. a powder balance for the private shooter), a set of grain or gram weights (pharmaceutical quality), a few beakers and measuring cylinders of convenient size, are about all the essentials. For those who systematically experiment, a burette or two with glass stopcocks and appropriate stand, are a convenience, but a few "dropping bottles" and a small glass syringe for dealing out small quantities of liquids not of strongly corrosive nature (such as concentrated nitric or sulphuric acids) are quite a useful substitute. For those yet more chemically interested, a set of small glass hydrometers, "giving densities from 0.8 to 1.5, range of each 0.1) for the approximate estimation and check of acid and similar concentration, is very recommendable. Density-concentration tables of acids, ammonia and many salt solutions are to be found in standard chemical notebooks and vade-mecums.

By taking the densities of the more commonly used liquid browne-components, volume-measurement can be substituted for the less convenient weighing of liquids, density being given by the simple relation "gram-weight divided by volume in ccm", and conversely, "volume in ccm = gram-weight divided by density". Running any given volume of a solution into a (tared) beaker and weighing, thus gives its density without the aid of a hydrometer.

4. Water

The same considerations as detailed under "Darkening," Chapter 11.5, apply here but with additional emphasis. Distilled water (obtainable at any druggist's) should invariably be used, as even filtered and boiled rain water almost always contains impurities, let alone accidentally absorbed acid.

5. Preparation

As a general rule, the solids are first dissolved, preferably in warm or hot water, the acid, alcohol, etc. added and after cooling the mixture brought to volume. With solutions containing bismuth, tin or antimony the acid must be added first in order to prevent the precipitation of basic salt, which dissolves with greater difficulty.

It is generally advisable to let the prepared solution stand for some time, in order to assure intimate mixing of its components and to allow unnoticeable minor internal reactions to settle out. This does not apply to such whose components set up reactions which sensibly modify their composition and effect progressively with age, until reaching a state of so-called "chemical equilibrium" or exhaustion of one of the main constituents. Such solutions are best prepared in small quantities and used after a certain uniform time following their preparation.

Solutions should always be kept in glass-stoppered bottles.

6. Chemicals

The most usual chemicals necessary for browning and similar gun purposes are here briefly described: in order to avoid all ambiguity as to meaning as far as possible, precise numerical data regarding strength, density, etc.—scarcely ever to be found in workshop and similar manuals—and in most cases the chemical formulae, are here given in each case.

Each chemical is here described under its English scientific name, popular trade designations and archaic names being added in English, French and German, also the scientific names in the latter languages when important to prevent misunderstanding, e.g. of two related compounds of the same metal. Sooner or later the more deeply interested shooter will want to experiment with foreign solutions, and the data here given will in many cases save him time and trouble.

Qualities. In general, chemicals of "commercial pure" (L. "purum venale" or "depuratum") are sufficiently good for browning purposes, and this should be specified in ordering from the drug store or chemical works: "analytical" quality is only occassionally necessary, and sometimes the "crude" will actually act more efficiently than the purer chemical.

Pharmaceutical preparations. Certain components frequently found in formulae for browning solutions are really medicinal preparations, most probably for no better reason than tradition and the fact that they are purchaseable at any local drug store. Such are for example "tincture of ferric chloride" and "sweet spirit of nitre." For medicinal purposes they must be prepared from the purest materials and according to strictly prescribed rule—different in each county and sometimes also at different periods. The precise origin and date of any given browning formula being mostly unknown, it is almost impossible to guarantee the absolute accuracy in composition when such components are in question.

The preparations referred to keep badly, and for browning purposes are advantageously replaced in the manner described under each.

Anhydrous and hydrated salts. Certain salts take up water on crystallising, mostly in simple molecular ratio, others form only anhydrous (water-free) crystals, others again crystallise in both forms according to surrounding conditions. As the water of crystallisation is of no value for the purpose intended, attention must be paid to this point in preparing solutions and in chemical calculations. A conversion table for such salts will be found at the end of the descriptions of chemicals. (Technical manuals are usually silent on this point).

Where the remark "cryst." follows the name of a salt in the formula for a browning solution, the hydrated, *not* the anhydrous salt is to be understood.

Densities. The densities here given are those of chemically pure substances: those of "commercial pure" quality therefore always differ slightly, so that the figures given are only an approximate measure of concentration, apart from the effect of temperature.

Acids will be considered first, as being a chief component of solutions: metric units being alone used in scientific descriptions, are the only ones given in this connection: table of equivalents on p. 149.

Hydrochloric acid, (L. acidum hydrochloricum, formula HCl, popularly spirits of salts, muriatic or oxymuriatic acid, whence correspondingly muriate or oxymuriate for "chloride:" Fr. "esprit de sel" "acid muriatique," G. "Salzsäure"), derives its name from its mode of manufacture (heating common salt with sulphuric acid), and is an aqueous solution of hydrochloric acid gas, having a sharp, penetrating smell, colourless when pure, but often yellowish from traces of iron, which at maximum concentration reaches a density 15/15 of 1.22, containing 43.09 G-% of gas.

In the U. S. A., France and Belgium the usual pure trade acid is one of D.1.19, containing 37.2 G-% of acid, or 44.3 g in 100 ccm. French "officinal" acid, D. 1.17, contains 33.46 G-%, or 39.1 g in 100 ccm. In Germany acids of both 1.19 and 1.16 density are available, whereas in England the latter is the usual trade quality, containing 31.5 G-% (average of data from various sources), or 36.6 g acid in 100 ccm.

Concentrated hydrochloric acid at the ordinary temperature is a strongly fuming liquid, giving off gas to atmosphere, consequently cannot be weighed without damaging the balance, of which more later. The strongest acid for ordinary purposes which at usual room temperature does not fume, has a content of about 20 G/V-%, obtainable by diluting 63.5 g=54.7 ccm of acid D. 1.16 or 53.8 g=46.2 ccm of D. 1.19 to a volume of 100 ccm.

Nitric acid, (L. acidum nitricum, formula HNO_3, popularly "aquafortis," Fr. "eau forte," G. "Scheidewasser") is manufactured similarly to hydrochloric acid, by heating nitre with sulphuric acid. Trade acids are of various qualities, of which the most usual are the following.

a. **Crude nitric acid,** (L. acidum nitricum crudum) is occasionally used for making up browning solutions: it is usually a yellowish to brown liquid, due to impurities and products of decomposition, of density 1.38 to 1.39, averaging 61 G-% of pure acid. On account of the presence of lower acid and even of chlorine, it acts more energetically than pure acid of equal concentration, and in certain cases, e.g. with difficulty oxidisable metals, is preferable to pure acid.

b. **"Commercial pure" nitric acid,** (L. acidum nitricum purum venale) is a colorless to slightly yellowish liquid of peculiar choking smell, the concentration most common in almost all countries being an acid of density 1.42, with a pure acid content of 69.8 G-%, equal to 99.1 g in 100 ccm.

Nitric is *the* oxidising acid par excellence, attacking nearly all common metals even when diluted, and with great violence when concentrated, except iron, and is one of the most frequent constituents of browning solutions. Exposed to air, it gives off biting fumes, absorbing moisture, although not very freely, decomposes slightly in bright sunlight, giving off reddish vapour (lower nitrogen oxides) and should therefore be kept away from direct light, to avoid ejection of the stopper. In diffused light the lower oxides are reabsorbed.

c. **Fuming nitric acids,** (L. acidum nitricum fumans) is a very concentrated, reddish-brown liquid containing a large amount of lower nitrogen oxides, which attains a maximum density of 1.55 and a maximum acid content of 99 G-%. The usual trade acid has a density round about 1.50, but this is no immediate guide to concentration, as the density varies irregularly with the degree of saturation with lower oxides. The following figures can serve as an approximate guide:

Density	G-% of acid
1.47	82.9
1.48	86.05
1.49	89.6
1.50	94.09

Fuming nitric acid is only occasionally used for browning, e. g. for quickly colouring small gun parts black, then used fully concentrated and as cold as possible: iron is then not attacked in the ordinary sense, but rendered "passive," on which effect its action is based.

On account of its content of lower nitrogen oxides, the diluted fuming acid—like the crude quality—is more strongly oxidising that the pure acid, and can thus be used to obtain a more active solution, in inverse proportion to their acid content.

As nitrous fumes are extremely deleterious, the fuming acid

must be handled with care, and whenever possible in the open or in well-ventilated rooms.

Aqua regia, (Fr. "eau régale," G. "Königswasser"), known since ancient times, is a mixture of concentrated hydrochloric and nitric acids, mostly in the volume-proportion of 3 to 4 parts of hydrochloric to 1 part of nitric acid. Its name is derived from its characteristic ability to dissolve the so-called "noble" metals, gold, platinum and its near relatives, which remain unattacked by nitric acid alone.

In essence aqua regia is a mixture evolving chlorine (by the oxidation of hydrochloric acid) which attacks all save a very few metals: the evolution of chlorine begins only after a certain interval following the mixing of the acids, and is recognisable by the smell and the small, rising gas bubbles: the aqua regia in then active.

The "nascent" chlorine ("in the act of formation") however sets up a series of secondary reactions, which go on side by side and consecutively until reaching the so-called state of "chemical equilibrium," or until the complete exhaustion of one of the original acids and their products of decomposition. The composition of all solutions containing both hydrochloric and nitric acids is therefore continually altering, finally losing their original active quality. Such solutions should thus be made up only in small quantities (e. g. 2 fl. oz.) shortly before use.

Theoretically only 2/3 of the chlorine content of the hydrochloric acid are directly available, the remainder going to form a so-called "mixed anhydride," called "nitrosyl chloride": this again reacts with water, and on account of the various secondary reactions which, according to concentration, temperature, presence of other constitutents and quantity of products of decomposition, take place in an unascertainable manner, it is difficult to estimate, even approximately, the real progress of these changes and their final result.

In the substitution—to be discussed in the sequel—of salts for the products of metal dissolved in aqua regia it is assumed as an approximate equivalent, that the amount of chlorine forming chloride is about the mean between that of the (imaginary assumed) "straight" and the above-described "aqua-regia-reaction," which, so far as the author's experience has extended, appears practically acceptable.

Such "salt-made" solutions differ externally from "acid-made" by being of lighter colour, the latter being darkened by the dissolved, reddish-brown nitrosyl chloride: this secondary product and its effects being absent, they are "duller" in action that the more active "acid-made" solutions and (quite generally) more stable than the latter.

To resume: solutions containing aqua regia, so long as active, are "keener" than single-acid solutions, are therefore (quite generally speaking) better adapted to the treatment of difficultly oxidisable metals: but the composition and physical state of the latter is of considerable influence, and the "keener" action may be of advantage or disadvantage according to the particular case. After completion of all internal reactions the aqua regia browne is at best as efficient as a single-acid one.

On the other hand, aqua regia solutions have the bedrock fault of a composition varying with the exact conditions of preparation, this and their effect continuously altering, unless prepared with strict method, that is, under repeated, identical conditions, and used always at the same subsequent age. In this way alone can comparable results be assured.

No lengthy reminder of the need for care in handling this mixture is needed, recalling the very deleterious nature of chlorine and nitrous fumes.

Sulphuric acid, (L. acidum sulfuricum, formula H_2SO_4, popularly "oil of vitriol," Fr. "huile de vitriol," G. "Vitriolöl"), is an infrequent ingredient of solutions, but extensively used for pickling and similar purposes. The dark-yellow to brown *crude* acid has mostly a density of 1.71, corresponding to a content of about 78 G-% of pure acid, or 133.3 g in 100 ccm. The "commercial pure" acid (L. acidum sulfuricum purum venale) is a syrupy, colourless to light-yellowish liquid of 1.84 density and 95 G-% acid content, equal to 174.8 g in 100 ccm.

Sulphuric acid is the chemically most powerful acid, and liberates all acids boiling at lower temperatures from their salts (examples under hydrochloric and nitric acids) and is used in industry for this and innumerable other purposes in enormous quantities.

The concentrated acid absorbs water, even when "combined," with great eagerness, attacking all organic substances with a violence which may go as far as complete carbonisation, and causes painful burns on the skin: caution therefore necessary. See further under "workroom acids" regarding dilution.

Phosphoric acid, (L. acidum phosphoricum, orthophosphoric acid, formula H_3PO_4, is only used for the production of rust-protective coatings ("phosphatising," Chapter VII). The regular American, English and German trade acid has a density of 1.75 and contains 88.9 G-% of acid, equal to 155.5 g in 100 ccm. The French trade acid has a density of 1.71, and a content of 85.8 G-%, equal to 146.7 g in 100 ccm: these are both colourless syrupy liquids. The acid according to Ph. Brit., mentioned in British patent specifications, has a density of 1.50, an acid content of 66.8 G-% = 100.2 g in 100 ccm, and is easy-flowing.

Workroom acids. All numerical data in solutions of acid content are to be uniformly understood as parts by *weight* of acid *in standard concentration,* namely of

Hydrochloric acid	D. 1.16.
Nitric acid	1.42.
Sulphuric acid	1.84.

Weighing small quantities of strong hydrochloric or nitric acid is to be avoided on account of their injurious fumes, that of strong sulphuric acid is troublesome by reason of its syrupy consistency, so that measuring by burette or measuring cylinder, according to the quantity required, is always preferable and sufficiently accurate for browning purposes.

These articles, obtainable at any chemical apparatus dealer's, are invariably graduated in metric units (universally used for scientific purposes), and acids convenient for workroom purposes are obtained by diluting to 100 ccm the following quantities and qualities:

	A. *Acid used*			*Quantity*	
				g.	ccm
	Hydrochloric	D.	1.16	64.8	55.8
	do.		1.19	54.9	46.1
	Nitric		1.42	64.8	45.6
	Sulphuric		1.84	64.8	35.2
B.	Hydrochloric	D.	1.16	50.0	43.1
	do.		1.19	42.4	35.6
	Nitric		1.42	50.0	35.2
	Sulphuric		1.84	50.0	27.2

Each cubic centimetre of diluted acid (A) then contains 10 grains by weight, and of acid (B) 0.5 gram by weight, of acid of standard concentration as defined above.

Acids thus prepared do not fume, are fluid, and each unit of volume bears the simplest possible relation to the quantities wanted in any particular case. They are best prepared in bulk and used in burette or cylinder as required.

Diluting. The *strict rule* in diluting strong acids is to run the concentrated acid *into* the water, *not* conversely, slowly and stirring continuously with a glass rod: this is particularly to be observed with nitric and sulphuric acids, which evolve great heat on dilution, and are liable to "spit," with extreme danger to eyes, if mixed the wrong way about. Diluting should be done in two stages, firstly to say 2/3 of the full extent, then after complete cooling brought to volume.

Alcohol, (L. alcohol aethylicus, ethyl alcohol, formula $C_2H_5.OH$, popularly "spirits of wine," Fr. "esprit de vin," G. "Weingeist, Spiritus") is a frequent component of browning solutions, in general too well-known to require description, but of very different qualities requiring more precise definition. In general, a good grain spirit is to be understood by this term, not potato- or burning spirit, or "denatured," that is, intentionally rendered unfit for human consumption by the addition of pyridine, kerosine, castor oil, etc. As a constituent of blueing, browning and other solutions the weight of "absolute," that is, 100% alcohol naturally alone counts: commercial "absolute" alcohol usually contains from 0.5 to 1% of water, which it absorbs with avidity. This by way of explanation, as absolute alcohol is never used in browning solutions. Alcohol concentration is usually expressed in volume-percentage (degrees Tralles).

Under "alcohol" in the sense of the U. S. and French Pharmacopoeias a 95° spirit is understood: in Belgium one of 94° is usual: that according to Ph.Brit. and G. is of 90° concentration, and these values apply to the "alcohol" occurring in brownes of corresponding origin (usually not known). Spirits of 70° and 56° are also listed in Ph.G. and French respectively.

Note on various alcohols.

"Potato spirit" is the commonest commercial alcohol, and is unsuitable for browning purposes: it contains amyl alcohol and many other impurities and can be distinguished by its unpleasant smell. Used for burning and common varnishes.

Alcohol "denatured" by the addition of methyl alcohol alone can be used (to advantage in England, where pure ethyl alcohol is extremely expensive on account of high Excise duty). *"Industrial methylated spirit,"* obtainable on special permit, contains 95 V-% of ethyl and 5 V-% of methyl alcohol, total alcoholic content 93.65 V-% or 90.53 G-%.

British "proof" spirit (so-called because when poured on gunpowder its water content just suffices to prevent the powder from igniting) contains 57.09 V-% or 49.3 G-% of pure spirit. Stronger or weaker spirits are called "over-" and "underproof": "25° overproof," means that 100 volumes of this spirit diluted with water yields 125 volumes of proof spirit, whereas "25° underproof" means one of which 100 volumes contain 75 volumes of proof spirit.

Alcohol "denatured" with kerosine and tarry matter, distilled over caustic potash and dehydrated over caustic lime is sufficiently purified for use instead of pure ethyl alcohol for most chemical experiments. (Illegal in the U. S. A. and England).

"Chloroform spirit" can be used as a substitute for pure ethyl alcohol in brownes, where the latter is exceptionally expensive or unobtainable. This contains 5 V-% of chloroform, which, being practically insoluble in water, is precipitated, consequently eliminated, on making up. It contains 84-87 V-% of alcohol, its maximum density being 0.865.

Note that the density of alcohol varies considerably with temperature, so that elaborate correction tables are needed for Excise and other purposes.

Apart from the cases in which 95° alcohol is positively specific, the alcoholic content of all solutions listed in Chapter IV is given in grains or grams of 90° spirit. For browning purposes the difference is practically negligible, and 10 grains of either can be taken as equivalent to a mean volume of 0.8 ccm, a gram occupying practically 1.25 ccm. The following tables of concentration and density, and for conversion of alcohols of different strength, will be found useful.

V-%	G-%	D. 15/15
100	100.00	0.794
95	92.40	0.816
94	91.04	0.820
90	85.70	0.834
70	62.46	0.890
56	48.24	0.922

A.   1 unit by weight of alcohol Column A is replaced by N units of strength

V-%	95°	94°	90°	70°	56°
100	1.082	1.098	1.167	1.601	2.073
95	1.000	1.015	1.078	1.480	1.915
94	0.985	1.000	1.062	1.548	1.887
90	0.927	0.941	1.000	1.372	1.777
70	0.676	0.686	0.729	1.000	1.295
56	0.522	0.530	0.563	0.772	1.000

Ether, (L. aether ethylicus, also popularly but incorrectly termed "sulphuric ether" in English, French and German), is an exceptional component of browning solutions. On account of its extreme volatility (b.p. abt. 35°C. = 95°F.) its value in solutions is problematical, and being in addition extremely inflammable it can easily cause accidents. The author has never been able to detect any difference in final result whether this constituent was included or omitted.

Ethyl nitrite solution, (L. spiritus aetheris nitrosi, spiritus nitri dulcis, spirit of nitrous ether, sweet spirit of nitre, Fr. "esprit de nitre dulcifié," "ether nitreux," G. "versüsster Salpetergeist," "Salpetrigsäureäther"), is a frequent component of solutions, especially in American and British formulae. It is really a pharmaceutical product, which must be prepared according to strict rule, and is a colourless, mobile alcoholic liquid of peculiar, agreeable smell and burning taste, containing from 1.5 to 4.5% of ethyl nitrite, a little aldehyde and traces of other organic bodies. The freshly-prepared spirit according to Ph.U.S.A. contains from 3.5 to 4.5%, to Ph.Brit. from 1.5 to 2.5%, to Ph.G. about 2% of ethyl nitrite. The solution is not very stable, and will only keep for a limited time, and according to the directions to be found in all Pharmacopoeias, should be kept in the dark in small, completely filled bottles.

For browning purposes it is advantageously replaced by the 15% spirit of nitrous ether (L. aether nitrosus 15%), which is a regular article of chemical trade in the U.S., France and Germany, and which, correspondingly diluted with alcohol, renders exactly the same service as the pharmaceutical article, at less cost and with the advantage of much greater stability. American chemical works also supply a very strong ethyl nitrite concentrate "1 to 21," which when diluted with 21 parts by weight of 95° alcohol produces 22 parts of a solution similar to the officinal one. (This concentrate is however liable to self-explosion in certain conditions.) In England a similar product called "spirit of nitre 7 to 1" is obtainable, containing from 16 to 20% of ethyl nitrite, and when diluted with 7 volumes of 90° alcohol is of approximately the same concentration as the officinal "spiritus aetheris nitrosi." It must however be noted that none of these substitutes is identical with the pharmaceutical preparation and must under no circumstances be used as such.

In Chapter IV this component of solutions is uniformly designated "spirit of nitre," with the indication, when known, of its percentage strength: where this is unknown a mean content of 2.5% can be taken as an acceptable average.

The density of "spirit of nitre" can be taken as sensibly equal to that of 90° alcohol, so that 10 grains by weight occupy nearly 0.8 ccm, and 1 gram very closely 1.2 ccm.

Ammonia, (L. liquor ammoniae caustici, aqua ammoniae,

formula $NH_4.OH$ or popularly abbreviated AmOH, "spirits of hartshorn," Fr. "alcali volatil," "esprit de sel ammoniaque," G. "Salmiakgeist," "Aetzammoniak"), is an important component of denickelling solutions, but is otherwise only occasionally used by the gunsmith for the immediate neutralisation of spilt acid.

In composition similar to that of hydrochloric acid, it is in contrast to the latter strongly basic: "ammonia" popularly so-called is an aqueous solution of ammonia gas, of well-known, extremely pungent smell, which can be considered as the hydroxide of the metal-resembling group NH_4, and thus counts as a base. In fact, it precipitates the oxides or hydroxides of all heavy metals from their salts, and has the special advantage over the "fixed alkalies" that all ammonium salt can be expelled from the residue by heat, thereby getting rid of any further disturbing influence.

Ammonia solution—more briefly "ammonia"—is lighter than water: the strongest commercial article has generally a density of 0.880 and contains about 35.5 G-% of ammonia gas: when warmed, gas is freely given off, the remaining solution becoming correspondingly weaker, and denser.

When converting formulae in English, it is to be noted that the "stronger ammonia" of Ph.U.S.A. has a density 15/15 of 0.901, containing 28 G-% of ammonia gas, whereas that of Ph.Brit. has a density 15/15 of 0.888 and an ammonia content of 32.5 G-%. Another concentration occasionally found in English recipes for denickelling is that of "ammonia 880" diluted with an equal volume of water, the density 15/15 of the mixture being 0.935 and the ammonia content about 17 G-%.

In spite of its basic character, ammonia is *unsuitable* for neutralising acid residues on metal parts, as its salts are mostly strongly rust-producing: if however it must be used for this purpose, all traces of its compounds must be at once removed by thorough rinsing with hot water.

Ammonium carbonate, (L. ammonium carbonicum, popularly salts of hartshorn, sal volatile, Fr. "sel volatil," G. "Hirschhornsalz," "flüchtiges Laugensalz"), is more correctly a mixture of ammonium carbonate and carbaminate, a salt of energetic basic reaction, smelling strongly of ammonia, which it moreover continuously gives off. It is used in denickelling.

Ammonium chloride, (L. ammonium chloratum, formula NH_4Cl, popular abbreviation AmCl, salammoniac, Fr. "sel ammoniaque," G. "Salmiak") the well-known rust-producing salt, is a frequent component of solutions, and can even be used alone as such with excellent results. (See Chapter IV, Group Aa. 2.)

Ammonium persulphate, (L. ammonium persulfuricum, formula $(NH_4)_2.S_2O_8$, abbreviated $Am_2S_2O_8$, is a strongly oxidising salt chiefly used for denickelling (See Chapter VI).

Ammonium sulphide, (L. ammonium sulfuratum), mostly used in solution (L. liquor ammonii hydrosulfurati), is a yellow, evil-smelling liquid of variable composition, used for retouching bright spots on browned parts ("Retouching," Chapter II.8). To be kept whenever possible away from direct light, in a cool place and in completely filled bottles.

The same purpose is served by so-called "liver of sulphur" (L. hepar sulfuris, Fr. "foie de soufre," G. "Schwefelleber"), a reddish-brown mass obtained by melting together potassium carbonate and sulphur, and consisting of mixed potassium sulphides.

Caustic potash and soda, (L. Kalium—or natrium—hydricum causticum, formulae KOH, NaOH, potassium or sodium hydroxide or hydrate, Fr. "potasse (soude) caustique," G. "Aetzkali" or "-natron") are only used by gunsmiths for neutralising or some other special purposes. Chemically, they are the strongest and most important bases, precipitating all other bases from their salt solutions, mostly as hydroxides, and are used to this end and for many other chemical and analytical purposes. Their effect is identical, but soda being cheaper is generally preferred.

Caustic alkalies strongly attack all organic matter—also the skin—and must therefore be handled with caution, the solids preferably with tongs, also keeping the hands away from strong solution: neutralise immediately if spilt on bench or gun-stock, with weak acid. The solid alkalies absorb atmospheric moisture and carbonic acid with avidity, so should be stored only in bottles with (slightly vaselined) ground glass stoppers.

Antimony trichloride, (L. stibium chloratum, formula $SbCl_3$, popularly but quite incorrectly called "butter of antimony" in English, French and German), is, on account of its use towards the end of the 18th century for browning the barrels of army guns (whence the name "Brown Bess") *the* classical browning medium.

Its old-time designation "butter of antimony" and its foreign equivalents, are quite loosely applied to quite a variety of substances, the muddle being only mitigated by the circumstance that the use of this salt is ever decreasing: the following explanations are added to clear matters up.

Antimony trichloride proper appears as translucent, slightly pinkish, anhydrous crystals, which gradually deliquesce and then further absorb moisture, attaining a maximum after about 70 days. On account of partial decomposition it then forms a turbid liquid containing 47.6% of anhydrous salt. When dissolved in water in the absence of free acid it deposits basic chloride.

"Butter of antimony," (L. butyrum antimonii, Fr. "beurre d'antimoine," G. "Spiessglanzbutter"), is strictly taken the correct designation of the deliquium of the chloride, that is, when just run to a soft, pinkish, pasty mass, then containing about 85% of anhydrous chloride.

Solution of antimony trichloride, (L. Liquor stibii chlorati, butyrum antimonii liquidum, G. "flüssige Spiessglanzbutter" of older German pharmacy—no longer listed in the newer editions of D.A.B.—and Ph.U.S.A. 8th, is a colourless to light yellow liquid (due to traces of iron) of density 1.35, which in 100 ccm contains 30 g of anhydrous chloride and 37.5 ccm of hydrochloric acid D.1.16.

Liquor stibii chlorati Ph.Brit. is obtained by boiling 47 g of purified antimony trisulphide (grey antimony glance) in concentrated hydrochloric acid and evaporation to a volume of 100 ccm. It is a reddish-yellow, strongly fuming liquid of density 1.47, containing 53.9% of anhydrous chloride plus free acid.

Boric acid, (L. acidum boricum, formula H_3BO_3, forms colourless, shiny, mother-of-pearl like scales or fine crystals, feeling somewhat soapy to the touch, and is a mild but thorough-acting derusting and matting medium. Details in Chapters VI and VII.

Bismuth chloride, more correctly oxychloride, (L. bismutum oxychloratum, formula BiOCl, G. "Perlweiss"), an anhydrous salt, component of certain particularly excellent browning solutions.

Bismuth nitrate, (L. bismutum nitricum neutrale, formula $Bi2NO_3,1aq.$) tried experimentally by the author in place of chloride, with good results (see Chapter IV, group Ac2).

> *Note.* In the absence of free acid all bismuth salts deposit basic precipitates: in preparing bismuthic brownes, the prescribed weight of acid should first be diluted with a little water, the

bismuth salt added and after complete solution the whole brought to volume.

In order to prevent precipitation with certainty, the chloride should be accompanied by a minimum of 5 times its weight of hydrochloric acid D.1.16, and the nitrate by at least 2.5 times its weight of nitric acid D.1.42.

Copper salts are a very frequent component of brownes, mostly as sulphate (L. cuprum sulfuricum, popularly blue vitriol or bluestone, Fr. "vitriol" or "couperose bleue," G. "Kupfer-vitriol"), more rarely as nitrate (L. cuprum nitricum) or chloride (L. cuprum chloratum), the choice between which is mostly governed by the nature of other components. They are all hydrated salts, their respective compositions being:

$$CuSO_4,5aq., \qquad Cu2NO_3,3aq., \qquad CuCl_2,2aq.$$

The sulphate forms large, hard, beautiful deep blue crystals, the two others smaller crystals, the deliquescent nitrate being beautiful dark blue, the chloride blueish-green.

Iron salts are found in the majority of brownes, both in the ferrous and ferric states, derived from their respective oxides.

Ferrous chloride, (L. ferrum chloratum oxydulatum, formula $FeCl_2,4$ aq., iron protochloride, Fr. "chlorure ferr*eux*," "protochlorure de fer," G. "Eisenchlor*ür*") is occasionally used. The chemically pure salt forms white, but generally light yellow to brown crystals, according to the degree of weathering, to which most ferrous salts are subject.

Ferrous sulphate, (L. ferrum sulfuricum oxydulatum, formula $FeSO_4,7$ aq., green vitriol, Fr. "sulfate ferr*eux*," "Vitriol vert," "couperose verte," G. "Ferrosulfat," "Eisenvitriol," "schwefelsaures Eisenoxyd*ul*;" see footnote) is by far the most-used ferrous salt: it forms large, clear green crystals, which easily oxidise on exposure ("weather") and become covered with brown hydroxide, so must be kept in airtight bottles.

The abominably misleading, not to say senseless English and German popular terms "copperas" and "Kupferwasser" cannot be too strongly condemned, the salt having no relation whatever to copper. In both cases the names have probably originated from the colour somewhat recalling verdigris, or by corruption of the French "couperose."

Ferric chloride, (L. ferrum chloratum oxydatum, ferrum sesquichloratum, iron perchloride, Fr. "chlorure ferrique," "perchlorure de fer," G. "Ferri-" or "Eisenchlorid"), crystallises both in anhydrous and hydrated form, but is mostly used in solution.

Anhydrous ferric chloride, (L. ferrum sesquichloratum anhydricum or sublimatum, flores Martis, formula Fe_2Cl_6, see footnote: Fr. "perchlorure de fer anhydre" or "sublimé," G.

"wasserfreies Eisenchlorid") appears as small hexagonal, purple-brown, intensely hygroscopic flakes, which absorb atmospheric moisture almost instantly, and therefore cannot be weighed accurately without special precautions. It deliquesces on exposure to a dark red, oily liquid, formerly known as "oil of Mars" (L. oleum Martis per deliquium, liquor stypticus Lofi, Fr. "huile de Mars," G. "Mars-" or "Eisenöl") having a density of about 1.75 and a content of anhydrous salt of about 66%.

The older formula is here retained in order to emphasize the relationship to ferric oxide.

Ferric chloride solution, (L. liquor ferri chlorati oxydati, or sesquichlorati. iron perchloride solution, Fr. "Perchlorure de fer dissous" or "liquide," G. "Eisenchloridlösung"), is a yet more frequent component of browning solutions than ferrous sulphate, and being obtainable ready-made from the chemical works or factor, is greatly to be preferred to the solid salt, which absorbs moisture while being weighed.

Unfortunately, the meaning of the term "ferric chloride solution" is very different according to country and date. as the following table will show:

| Country. | Date. | Percentage by weight of | | Density 15/15 |
		Ferric	chloride.	Metallic iron.	
		anh.	hydr.		
England. .... (1).	1914.	13.00	21.67	4.47	1.114
do.	1898.	14.40	24.00	4.96	1.127
France.	1910.	26.00	43.33	8.95	1.247
Germany. ...... (2).	1910.	29.05	48.43	10.00	1.282
U S. A.	1890.	37.76	62.93	13.00	1.388
England. .... (3).	1914.	40.75	67.92	14.03	1.426
Germany.	1890.	43.58	72.63	15.00	1.462
Holland.	1905.	do.	do.	do.	do.
England. .... (3).	1898.	44.40	74.00	15.28	1.473
Italy. .... (4).	1892.	44.50	74.17	15.32	1.477

(1). On account of the (acknowledged) inaccuracy of densities given in Ph.Brit., all data regarding British solutions are only approximate.

(2). Also U. S. A. 1908 to 1936, Austria 1906, Belgium 1906, Hungary 1909, Japan 1907, Russia 1911 and Switzerland 1904.

(3). "British strong solution" (L. liquor ferri perchloridi fortis).

(4). Average.

The densities of the above table are taken from Hager's "Handbuch der Pharmceutischen Praxis" Berlin, 1900, reduced to 15°C. no two density tables to be found in chemical literature agreeing.

In the absence of precise information as to the origin and date of a solution containing ferric chloride, the far most frequently-occurring concentration according to Ph.U.S.A. 1908 and onwards may be assumed, and in all browning formulae given in the present work is listed as "sol. ferric chloride 29%."

In contact with iron it is at once reduced to ferrous salt, dissolving a certain amount of metal: the ferro salt is in turn oxidised on exposure to ferric hydroxide, and so on, thereby effecting the desired oxidation of the surface.

Hydrated ferric chloride, (L. Ferrum sesquichloratum hydricum cryst., chloretum ferricum, Fr. "perchlorure de fer cristallisé," G. "Ferrichloridhydrat" or briefly "Eisenchlorid"), formula $Fe_2Cl_6,12aq.$ but containing traces of higher hydrates, is the most important ferric salt, when chemically pure containing exactly 60% of anhydrous chloride, or 20.66% of metallic iron. It forms a yellow, crystalline mass, still very hygroscopic, although very much less so than the anhydrous chloride.

Where solid ferric chloride is mentioned in the sequel, this hydrate, not the anhydrous salt, is meant.

"Tincture" of ferric chloride, (L. tinctura ferri chlorati oxydati, tinctura ferri perchloridi, tincture of steel, Fr. "teinture de fer" or "d'acier," G. "Chloreisentinktur," "Stahltinktur," "Stahlfarbe"), is nothing more than an alcoholised solution of ferric chloride ("tincture" in pharmacy means an alcoholic solution), is a component much favoured in American and British browning formulae, to what good end is difficult to understand. It keeps badly, decomposing after relatively brief storage, especially if exposed to light and air, with formation of deposit.

Like ferric chloride solution, "tincture" means different things in different countries and at different dates: the "tincture" of the newer editions of Ph.G. (D.A.B. V and VI.) also contains ether ("tinctura aetherea ferri sesquichlorati"), but so far as concerns browning, this component may as well be omitted. It can be safely assumed—in default of evidence to the contrary —that the majority of formulae containing "tincture" to be found in German sources, are of older date: in the sequel, and so far as more recent origin cannot be ascertained, the composition of Ph.G 1890 has been assumed.

Briefly, "tincture," with its bad keeping qualities, is without useful object in the gunsmith's chemical cupboard, and whenever found in formulae, is advantageously replaced by its components when making up solutions, according to the following table, in which the ether of "tincture" according to D.A.B. V and VI is replaced by alcohol.

Pharmacopoeia.	1 unit by weight of official "tincture" is replaced by N unit of		
	Sol. ferric and chloride 29%	Alcohol or 90°	Alcohol 95°
Brit. 1898.	0.50	0.20	0.19
G. III.	0.20	0.60	0.55
D.A.B. V., VI.	0.10	0.90	0.85
Fr. 1920.	0.10	0.50	0.45
U.S.A. Rev. 8 to 11.	0.45	0.60	0.55

All "tincture"—containing formulae reduced in the present work to a uniform basis have been thus dealt with, which explains the occasionally very small (in many cases negligible) amounts of alcohol appearing. Apparently only its more "learned"-sounding name accounts for its inclusion.

Ferric nitrate, (L. ferrum nitricum oxydatum, formula $Fe_2 6NO_3$, 18 aq., iron pernitrate, dissolved iron oxide, Fr. "nitrate ferrique," "oxyde de fer dissous," G. "Ferri-" or "Eisennitrat," "Salpetersaures Eisenoxyd"), is less generally used in solutions than the chloride, and is less strongly rust-producing in equal concentration, but gives equally fine results as to finish as the chloride.

Ferric nitrate appears as rather large, light violet-coloured, transparent crystals, smelling faintly of nitric acid, which easily deliquesce to a brown liquid. Its water-content varies according to the conditions accompanying crystallisation, and the salt is therefore best ordered in "analytical" quality, which has the composition given above. Instead of the solid salt, the trade solution "liquor ferri nitrici oxydati D.1.25" (e. g. Merck's) can be used: nitrate-content conflicting in literature, but according to determinations made by the author a mean of 28 G-% of anhydrous salt, that is, 35.0 g of anhydrous or 58.5 g of "analytical" nitrate in 100 ccm.

Now and again the archaic designation "flüssiges Eisenoxyd" is to be found in German, Austrian and Swiss browning formulae, which, after some searching, was discovered to mean a ferric nitrate solution containing 5 G-% of metallic iron, that is, 21.66 G-% of anhydrous salt. It can be obtained by diluting 77.5 g = 62.0 ccm of the 28% solution to a volume of 100 ccm.

Ferric sulphate, (L. ferrum sulfuricum oxydatum, iron persulphate, Fr. "persulfate de fer"—both incorrect—correctly "sulfate ferrique," G. "Ferrisulfat," "Eisensesquisulfat,"

"Schwefelsaures Eisenoxyd"), is the ferric salt of sulphuric, not *per*sulphuric acid.   In contrast to ferrous sulphate it is a stable, anhydrous or hydrated salt.

Anhydrous ferric sulphate (L. ferrum sulfuricum oxydatum siccum or anhydricum, formula $Fe_2(SO_4)_3$, is a white, very hygroscopic mass, running to a brown syrup on exposure, and is not a common product.

The hydrated trade salt is of variable water-content, and at different times the author has had lots varying between 18 and 32% of crystal-water: when ordering it is best to ascertain the exact content of anhydrous salt from the factory or maker, or to determine it at home when received (*).

> (*). By igniting (strongly heating) 10 to 15 g (say 150 to 225 grs.) of salt in a covered porcelain or silica crucible.   If (A) is the weight taken, and (a) that of the residue after ignition, $\dfrac{250\ a}{A}$ is the percentage of anhydrous sulphate in the sample.

Ferric sulphate solution, (L. liquor ferri sulfurici oxydati, liquor ferri persulfatis or tersulfatis, Fr. "persulfate de fer liquide," G. "flüssiges schwefelsaures Eisenoxyd"), is a clear, rather syrupy, brownish-yellow to reddish-brown liquid, which according to Ph.U.S.A., Brit. and Swiss contains an average of 10 G-% metallic iron, or 27.9 G-% of anhydrous salt, plus free sulphuric acid, and according to the amount of the latter has a D.15/15 of 1.428 to 1.441.   If used for browning, first ascertain its acid content and allow for accordingly.

Ferric sulphate is not frequently found in browning solutions, and so far as the author's (not very numerous) trials with it have gone, it would appear to have no appreciable advantage over the chloride or nitrate.

Mercuric chloride, (L. hydrargyrum bichloratum or muriaticum corrosivum, formula $HgCl_2$, corrosive sublimate, muriate or oxymuriate of mercury, Fr. 'chlorure mercur*ique*," "sublimé corrosif," G. "Quecksilber" or "Mercurichlor*id*," "Aetzsublimat"), is a well-known, anhydrous, strongly rust-producing body, of frequent occurrence in browning solutions.   At ordinary temperature it is soluble to the extent of about 6% in water and 22% in alcohol.

Being a powerful poison, it is important to avoid confusing it with the insoluble mercur*ous* chloride, better known as "calomel," (L. hydrargyrum chloratum mite, mercurius dulcis or mitigatus, Fr. "chlorure mercur*eux*," G. "Quecksilberchlor*ür*"), valuable for medicinal purposes.

Mercurous nitrate, (L. hydrargyrum nitricum *oxydul*atum, formula $(Hg)_2NO_3$,laq., Fr. "nitrate mercur*eux*," G. Mercuronitrat," "salpetersaures Quecksilber*oxydul*"), is only used in the composition of certain etching solutions for particular qualities of steel.

Mercuric sulphate, (L. hydrargyrum sulfuricum oxyd*atum* neutrale or bisulfuricum, formula $HgSO_4$, Fr. "sulfate mercu*rique*," G. "Mercu*ri*-" or "Quecksilbersulfat," "schwefelsaures Quecksilberoxyd"), exceptionally used, less soluble and oxidising than chloride, appears of doubtful value for solutions.

Note. In the absence of free acid, these two salts are decomposed by much water, therefore add acid first.

Potassium and sodium carbonate, (L. Kalium or natrium carbonicum), formulae K_2CO_3, Na_2CO_3, common potash and washing soda, are used by the gunsmith for degreasing, less frequently for neutralising. Although "normal" salts, they react strongly alkaline, which explains itself by the combination of a strong alkali with a weak acid.

"Potash" is anhydrous but deliquescent, "soda" hydrated (Na_2CO_3,10aq.) but efflorescent on exposure and therefore to be preferred: effect otherwise the same. The popular expressions "sal-soda" and "soda-ash" respectively mean ordinary washing- and anhydrous (Solvay) soda and are effective in almost the exact proportion of 8 to 3 parts by weight.

Potassium bisulphate, (L. kalium bisulfuricum or sulfuricum acidum, formula $KHSO_4$, acid potassium sulphate, Fr. "bisulfate de potasse," G. "Kaliumhydrosulfat," "saures Kaliumsulfat"), is used for matting and derusting: details in Chapter VI.

Potassium chlorate, (L. kalium chloricum, formula $KClO_3$, oxymuriate of potassium, Fr. "oxymuriate de potasse," G. "chlorsaures Kali") and nitrate (L. kalium nitricum, formula KNO_3, nitre, Fr. "salpètre," G. "Kalisalpeter" or shortly "Salpeter") are anhydrous, small brilliant crystals of cooling taste, used in certain "express" browning solutions and fluxes. *Sodium* nitrate ("Chili saltpeter") was until the development of the synthesis of ammonia from atmospheric nitrogen, so to speak the sole basic product for nitric acid manufacture.

Potassium dichromate, (L. kalium bichromicum, formula $K_2Cr_2O_7$, Fr. "bichromate de potasse," G. "doppelt chromsaures Kali"), generally but less correctly known as "bichromate," appears as large, anhydrous, hard orange-coloured crystals, likewise used in a few solutions and for fluxes: less oxidising than chlorate.

Tin chlorides are seldom used in browning solutions, but chiefly for derusting: there are two of these, which must be carefully distinguished.

Stannous chloride, (L. stannum chloratum, formula $SnCl_2$, 2aq., commonly known as "tin salt," Fr. "sel d'etain," "chlorure stann*eux*," G. "Zinnchlor*ür*," "Zinnsalz"), forms white crystals, soluble in water acidified with hydrochloric acid: in the absence of acid an oxychloride is immediately precipitated.

Stannic chloride exists in both anhydrous and hydrated forms, very different physically.

Anhydrous stannic chloride,   (L. stannum bichloratum fumans, spiritus fumans Libavii, formula $SnCl_4$, Fr. "chlorure stann*ique*," G. "Zinnchlor*id*"), is a heavy, colourless, very mobile and strongly refracting liquid of density 2.23, fuming strongly on exposure, by absorption of moisture, strongly attacking eyes and lungs, an unpleasant body to handle.

Mixed with 1/3 of its weight of water, it forms, accompanied by hissing and great evolution of heat, a soft, white to grey paste known as "butter of tin" (L. butyrum stanni, Fr. "beurre d'etain," G. "Zinnbutter"), which solidifies to a crystalline mass on cooling, and again softens when gently warmed. If it is desired to use the anhydrous chloride as the point of departure, it is best converted from the start into the "butter," which can be handled without inconvenience.

Hydrated stannic chloride, (L. Stannum bichloratum hydricum, formula $SnCl_4$,5aq., Fr. "chlorure stannique hydrate," G. "Zinnchloridhydrat"), is the easiest-obtainable commercial hydrate, a white to greyish salt of almost identical composition to "butter of tin," the use of which in place of the anhydrous chloride can only be recommended.

Zinc chloride, (L. zincum chloratum, formula $ZnCl_2$, popularly "killed spirits," Fr. "chlorure de zinc," G. "Zinkchlorid," "Chlorzink") an occasional component of browning solutions, is a strongly rust-producing salt of very acid reaction.   The trade article appears in the form of thin cast sticks, extremely hygroscopic, which run after a short exposure to a syrupy liquid: as it strongly attacks the skin, it should be handled only with tongs or similar instrument.

Mixed with $\frac{1}{4}$ its weight of water, it forms a soft, pasty mass known as "butter of zinc" (L. butyrum zinci, Fr. "beurre de zinc," G. "Zinkbutter"), a term often but inaccurately applied to the chloride itself.

Solution of zinc chloride, (L. liquor zinci chloridi), according to Ph.U.S.A., Brit. and G. contains, within very small variations 50 G-% of chloride, is a heavy liquid of density 15/15 of 1.565, the use of which, or of "zinc butter" is to be preferred to that of the corrosive solid chloride.

Zinc sulphate, (L. zincum sulfuricum, formula $ZnSO_4$,7aq., white vitriol, Fr. "couperose blanche," G. "weisser Gallizienstein"), is an efflorescent salt, now and again occurring in solutions, and much less keenly rust-producing than the chloride.

More rarely used chemicals are briefly described under the headings in which they occur.

Special grease solvents.   The properties of those mentioned in Chapter II will be here briefly described.

Carbon disulphide, (L. carboneum sulfuratum, alcohol sulfuris, formula CS_2, Fr. "sulfure de carbone," G. "Schwefelkohlenstoff"), is a colourless to yellowish, mobile, strongly refracting (crystalline-bright) liquid, D. 15/15 1.26, b.p. 46°C. (abt. 110°F.). When pure it has a peculiar but not unpleasant aromatic smell, but when exposed to light and especially in presence of moisture it becomes evil-smelling: crude carbon disulphide is appallingly foetid and produces retching. It should therefore be preferably kept in the dark.

Carbon disulphide is the most effective eliminator of the last traces of grease from gun parts, but by reason of its low boiling point (the sun temperature of a summer's day) and its extreme inflammability is unfortunately extremely fire-dangerous: its heavy vapour sinks to and creeps along floors and mixed with air explodes violently in contact with any flame.

In addition to its stupifying and and anaesthetising qualities carbon disulphide is a powerful nerve poison, producing on prolonged inhalation of its vapours very dangerous organic commotions: it should therefore be used only in the open or in particularly well-ventilated rooms.

Carbon tetrachloride, (L. carboneum tetrachloratum, formula CCl_4, tetrachlormethane, Fr. "tetrachlorure de carbone," G. "Tetrachlorkohlenstoff"), and **trichlorethylene,** (L. aethylenum trichloratum, formula C_2HCl_3), density respectively 1.59 and 1.47, b.p. 77°C. and 87°C., are heavy, colourless, less strongly refracting, non-inflammable liquids of agreeable smell resembling that of chloroform (to which they are chemically closely related), very excellent grease solvents, although not quite so keen as carbon disulphide, but offering the advantage of complete safety from fire-raising.

Both are distinctly narcotic and nerve poisons, but being less volatile than carbon disulphide are less dangerous in this respect.

Chloroform itself is also an excellent grease solvent, but for obvious reasons its use for this purpose is not to be recommended.

Note. As the above particularly effective grease solvents are much more expensive than gasoline, their use—except where the scale of operation justifies the cost of a recovery plant—is only commercially possible exceptionally and for small parts.

Conversion table for salts. The following table gives the conversion factors, in terms of weight, for the reduction of hydrated to (effective) anhydrous salt, and vice versa.

Salt.	Anhydrous. Hydrated.	Hydrated. Anhydrous.
Ferrous chloride	0.637	1.588
Ferrous sulphate	0.546	1.830
Ferric chloride	0.600	1.667
Ferric nitrate	0.599	1.670
Copper chloride	0.788	1.306
Copper nitrate	0.777	1.288
Copper sulphate	0.639	1.563
Mercurous nitrate	0.967	1.035
Bismuth nitrate neutr.	0.815	1.228
Stannous chloride	0.841	1.190
Stannic chloride	0.742	1.345
"Butter of tin" [1]	0.750	1.333
Zinc sulphate	0.561	1.780
"Zinc butter" [1]	0.800	1.250

[1] Standard preparation.

7. Storing Chemicals

Certain hydrated salts, e. g. common washing soda, gradually lose their water of crystallisation on exposure: such are known as "efflorescent." Others gradually absorb moisture, finally running to liquid, and are thence called "hygroscopic" or "deliquescent." As both actions continually change their composition, and certain salts attack metal, all salts should always be kept in glass-stoppered bottles; slightly vaselining stoppers and the inside of necks will ensure complete airtightness and prevent the stoppers from sticking. Tins are mostly quite unsuitable.

If not vaselined, stoppers occasionally stick, and can be loosened in various ways: one of the best is to gently warm the neck of bottle over a burner turned low, or a candle flame, continually turning so as to warm the neck uniformly, and in a very short time the stopper can be easily removed. Care must be taken to let the neck cool completely before replacing the stopper.

With inflammable contents this process is obviously risky, as there is no guarantee against sudden cracking: if then the bottle is held firmly by one person, a turn of fairly thick string taken round the neck and pulled briskly to and fro under slight tension, the friction will presently warm the neck sufficiently to enable the stopper, assisted if necessary by gently tapping to and fro, to be extracted.

Another way in such cases is to pour warm, hot and boiling water in succession on the neck, turning continually, and taking care to avoid contact of contained liquid with the heated neck, which may easily cause cracking. For this reason the string method is preferable.

The stoppers of bottles containing concentrated salt solutions occasionally get formally cemented in the neck if left long un-opened, resisting all methods of loosening, in which case the drastic step of breaking the bottle is the only resource. A more generous allowance of vaseline in neck and on stopper is the best preventive.

PART II
Special Division

CHAPTER IV

BROWNING SOLUTIONS AND WORKING INSTRUCTIONS

General Considerations

The composition of browning solutions was jealously guarded by the old-time gunsmiths as a trade secret, but at the present day there is only the embarrassment of choice, and those in the present work—over 200—collected by the author in the course of years, are probably only a fraction of all existing ones. Certain of these are without doubt better than others, but the extravagant claims made for some new (or resurrected) ones advertised are by no means to be taken at face value. As already emphasized, irreproachable "brown" finish can be gotten with the very simplest means, and there is a good deal more "boost" than solid merit in the more complicated formulae.

Serious reproach can be levelled at the literature of the subject on the score of vagueness: precise statement of acid strength, percentage of effective substance in solutions, the general influence of this or that component, and definite working instructions, are the exception rather than the rule. In addition, use is frequently made of unusual, local or out-of-date names of ingredients, the meaning of which cannot be found in any technological encyclopaedia.

The formulae here collected being culled from sources of different nationality, the first task was to ascertain, as far as possible the corresponding meaning of each substance-name. This was by no means facilitated by the circumstance that nothing proved that a formula appearing for example, in English, did not originate in Belgium or Austria, etc., where the same name meant something sensibly different than in America or England. The confusion of formulae appearing in English is increased by the large variety of Anglo-American weights and measures, and the differences between units of the same name in the two countries.

Although all possible trouble has here been taken to complete deficiencies and to replace vagueness by accuracy, it will be easily understood that the compositions here given, reduced to a uniform basis, can only claim comparative and not absolute precision.

In the literature of the subject, a reason for the inclusion of this or that ingredient is rarely to be found: the technique of browning has developed quite empirically, and at best a few, rather uncertain rules are recognisable, subject however to considerable exceptions: the widely differing concentrations can be cited in proof. The following general considerations resulting from the experience of others and the author, are offered as a guide.

Concentration. Concentrated browning solutions, as may be expected, act more rapidly, diluted ones more slowly and less keenly: according to the old, accepted rule, the latter produce coatings of closer grain, better appearance and greater durability: here the practice of luxury gunmakers can be taken as evidence. As metals are not absolutely homogeneous, concentrated browning solutions are more liable to act unevenly, owing to more intensive local action, whence the old-time practice of weaker brownes for twist barrels.

Acids. Most brownes contain either nitric or hydrochloric acid, a few both together, that is, aqua regia, usually to be avoided (see under this heading, p. 36). Hydrochloric browning solutions mostly act more rapidly than nitric, but incline more to after-rusting, but other factors, such as concentration, composition of metal, presence of other salts, also intervene. Sulphuric acid is less frequently used and acts less energetically than hydrochloric or nitric.

Metallic salts are best obtained ready-made from the chemical works or dealer. Many formulae for browning solutions give such combinations as "iron filings and nitric acid," "zinc carbonate and hydrochloric acid," which only involve more trouble without the least compensating advantage, and are at best only of service in the absence or difficulty of obtaining the salt, but immediate availability of its components (e. g. in the U. S. A. on account of the continental distances). Moreover, the evolution of corrosive, deleterious or even poisonous gases sometimes involved, by no means adds to the pleasure of home preparation.

Only in cases—e. g. where solutions include aqua regia—and a "clear" reaction cannot be guaranteed, is the original as well as the estimated substitute composition given in the sequel.

Chlorides are the most frequently used salts, sulphates less, and nitrates only now and again: iron (ferrous) and copper sulphates probably because these are the most commonly procurable salts of their metals.

Heavy metal salts, except those of zinc, brought into contact with iron, immediately form on it a precipitate of their base metal: apart from the purely chemical effect, thereby forming a galvanic couple which promotes rusting of the electropositive iron. The precipitates also become oxidised in the succeeding operations and thus influence to a certain extent the colour-tone of the final finish.

Ammonium, ferric, mercuric and zinc chlorides are specially energetic rust-producers: browning solutions consisting mainly of ferric chloride and hydrochloric acid incline to produce finishes shading into yellowish- to reddish-brown (Brandeis). Those containing antimony trichloride give various tones according to composition (see under this heading, p. 43).

Excessive copper-content easily conduces to coatings that readily peel off: excess of ferric chloride is equally harmful, as it partly redissolves the rust already formed, causing roughened surface and unsightly spots, not to say pits, due to over-energetic action.

Brandeis ("Die moderne Gewehrfabrikation") rightly emphasizes the specially favourable effect of ammonium chloride, remarking that its use produces as deep a black as is possibly obtainable by any means (see under "Salammoniac brown, Group Aa, p. 62).

Alcohol, generally speaking, retards action, therefore assists in producing a coating of close grain, more strongly adherent, and of more brilliant finish, and is thus of real advantage. Alcohol does react with nitric acid when also present, but in the usual degrees of concentration unnoticeably and without detriment to the general effect of the solution.

Ethyl nitrite ("spirit of nitre"). This body, alone or with further addition of alcohol, is a very frequent component of browning solutions, especially those of American or English origin, whether on account of real merit or tradition, is not immediately determinable. According to legend, it assists, like alcohol and ether, in promoting closeness of grain, greater brilliancy and better adherence in the coatings: it may fairly be questioned whether this effect is not due to the far larger quantity, relatively, of the alcohol also present.

Ethyl nitrite is in itself no very stable body: as an "ester" (organic equivalent of the mineral "salt") it saponifies (splits up) easily and readily oxidises to nitrate, actions which are promoted by warmth, bright light and the presence of oxidising substances. (Whence the admonition of Ph.U.S.A. and other, to store in a cool and dark place.)

On this account (and under reservation of the considerations to follow) the author is inclined to believe that the value of ethyl nitrite in solutions has often been over-estimated: in a whole series of comparative trials he has been unable to detect any superiority in respect of final result with a browne containing ethyl nitrite, compared to one with ethyl nitrate or plain alcohol.

Ethyl nitrite in hydrochloric solutions. The above applies particularly to hydrochloric brownes: the concentrated acid does react with ethyl nitrite, discolouring it, but in the usual concentrations prevailing in brownes this action remains unnoticeable and without ill-effect on the final result.

Ethyl nitrite in nitric solutions. Things are materially different in presence of nitric acid and metallic salts: in certain brownes containing these components an unmistakeable reaction, recognisable by the rise of small gas bubbles and the characteristic smell of nitrous fumes, sets in after a certain time, pointing to the oxidation of the nitrite, at the same time producing nitrous acid. Bright sunshine specially promotes this action, and its completion is recognised by the disappearance of the peculiar corrosive smell, the cessation of gas evolution, and a change in colour of the browne.

As the presence of nitrous acid—especially in the "nascent" state, that is, during continuance of the reaction—increases the oxidising intensity of nitric acid (see under "crude" or "fuming" nitric acid, p. 34), a solution in this state of "fermentation" (figuratively put) is keener-acting than ordinarily, and on completion relapses into the normal activity of a partially-exhausted "plain" browne. The ethyl nitrate formed is not further attacked, and its subsequent effect is at best slightly retarding, like that of alcohol. The author, much interested in these effects, has been unable to recognise any difference in the quality of the finish produced by spent ethyl nitrite browning solution, except that a fourth processing is occasionally necessary or desirable.

To resume: the value of ethyl nitrite in a nitric browning solution would appear mostly to consist, in certain cases, in a real but temporary "activation" of the browne, but before this observed effect can be accepted as a general rule, it requires confirmation—or refutation—by others and longer comparative trial. Finally, "keener"-acting brownes without ethyl nitrite or secondary reactions, can be obtained by the simple substitution of "crude" or "fuming" for plain nitric acid.

"Express" browning solutions. Finally, the author would mention that he obtained better results with brownes of this class (Groups Ac. and Ad.), whether containing or not containing acid, by leaving out the ethyl nitrite altogether: in the presence of such a strong oxidant as chlorate the real utility of the nitrite addition may well be questioned.

American, British and Metric
Weights and Measures

The complication of American and British weights (avoirdupois, apothecaries' and Troy) and the differences between American and British measures of the same name often render the correct interpretation of browning formulae appearing in English a matter of uncertainty, and the absence of decimal connection does not facilitate transition from one quantity to another in a different unit.

Passing over details, it is important to note that whereas the U. S. pint, quart and gallon (the old so-called "Winchester" or "wine" measure) are *smaller* than the British, the U. S. fluid ounce is *larger*, the pint being differently subdivided in the two countries.

Metric units always mean the same thing, and being decimally interconnected are in universal use for scientific purposes.

The present work being intended chiefly for the individual gun owner and smaller trade gunsmith, all browning formulae are given in both systems, on the uniform basis of 1/4 U. S. pint and 100 ccm respectively, the weights being given in grains (the only unit common to all Anglo-American varieties) and grams, to suit those possessing sets of either.

To convert to and from metric: the "1/4 U. S. pint-grain" weights multiplied by 0.0548 give the "100 ccm-gram" weights, and conversely, the latter multiplied by 18.255 give the "1/4 U. S. pint-grain" values. (For *British* 1/4 pint quantity—5 British fluid ounces add 1/5 to the 1/4 U. S. pint-grain weight of each chemical—the corresponding conversion factors are: *from* metric, 21.926, and *to* metric, 0.0456. All are easily applied by the slide-rule).

German browning formulae of older origin are occasionally found in the old "Nuremberg Medicinal and Apothecaries'" units: a table of the metric equivalents of these as well as of American and British units will be found on p. 150.

Where reference is made to "drops," the value of 0.06 ccm adopted by International Convention can be taken: using a drop tube 3 mm in external diameter, 20 drops of distilled water delivered under the minimum head at a temperature of 15.5°C. (60°F.) make a volume of exactly 1 ccm.

Classification of Browning Solutions

The special steels increasingly used for gun barrels and parts vary considerably in composition, and it can easily happen that a solution giving excellent results on a certain kind of steel gives poor or even negative results on another. So far as the author has been able to discover, no systematic research has yet been carried out, enabling the composition of a solution best suited to a steel of definite composition to be determined in advance.

In order to present the very numerous formulae here assembled according to an orderly plan, they have been arranged according to their composition: this classification, which the author believes to be the first ever proposed, will give useful service for use and reference until a better one, based on the results given by each solution, has been evolved.

Brownes thus fall into definite groups, as follows:

```
Group Aa. Iron-free, containing 1 metallic salt.
       Ab.       "            "      2 metallic salts.
       Ac.       "            "      3    "      "
       Ad.       "            "      4    "      "
                 "Express" brownes.  Troubles.  Critical remarks.
Group B.  Containing iron, no other metallic salt.
       C.        "         "    and 1      "        "
       D.        "         "     "  2 other metallic salts.
       E.        "         "     "  3    "        "      "
Group F.  Combined brownes.
       G.  Collective browning.
       H.  Miscellaneous browning and blueing methods.
       K.  Browning "Antinit" and "Anticorro" Steel.
```

Personal experiences is indicated by smaller type: where no working instructions follow the composition of a solution, it has not been tried out by the author.

"Room temperature" means one between 15° and 20°C. (abt. 60° to 70°F.), "steam oven temperature" one of 35° to 40°C. (95° to 105°F.) where not otherwise mentioned.

Comparative trial. In order to form a preliminary estimate of the value of a solution, the result of a comparative trial of a certain number, made under identical conditions, is given hereafter: these conditions being

> "Natural" rusting at room temperature.
> Three passes (rustings).
> Boiling and scratching between each.
> Trial pieces: small flat pieces of machine and tool quality carbon steel, as purchased.

The results given are simply intended to provide a term of comparative excellence, but in no wise to be understood as the best the solution can give under optimum application.

> *Note.* All browning solutions were converted to metric units and standard volume as and when found: on account of the small corrections needed to obtain "round" figures, they may not exactly reproduce the original when reconverted, but the very small divergences are without practical importance.
>
> The references given simply indicate where the particular formula was first found, and do not necessarily represent their absolute origin, mostly unknown.

GROUP Aa

Iron-free Brownes with 1 Metallic Salt (*)

Aa. 1. Browning with common salt.

The rusting effect of salt water has been known since the beginning of the iron age, and it can at least be surmised that common salt was the first substance intentionally used to produce rust. So much at least is certain, that it was in common use for browning gun barrels, from about the beginning to the middle of the 18th century.

As a matter of historical interest, the author made a few trials with salt, without suggesting that the following was the exact method used by the early gunsmiths.

> a. Without boiling. Fairly strong (8 to 12%) salt solutions were used, by which a brown, strongly-adherent rust was slowly formed at room temperature. Thorough scratching was a work of time and patience, especially with wrought iron, which has not the homogeneity of modern ingot iron and steel.
>
> The final result after 8 to 15 passes was a greyish-brown, earthy-looking coating, whose appearance was improved by repeated rinsing with (or ½ hour's soaking in) very weak (abt. 0.2%) copper sulphate solution, the colour gradually

---

* The group Ammonium, NH₄, is here assimilated to a metal.

shading into plum-brown by the deposit of metallic copper, further assisted by steeping for ½ hour in hot decoction of logwood (2 g per litre = 30 grs. per U. S. quart): adding abt. 7.5 grs. of ferrous sulphate the coating is rendered almost black.

Then rubbing with linseed oil or wax, or preferably lacquering.

b. **With boiling.** More from curiosity than for intended use, the author tried the procedure of the comparative trial, boiling and scratching after each pass. From 3 to 4 passes then gave a dark-brownish, very acceptable black, not requiring lacquering and presenting a very passable appearance.

So that when nothing else is available, common salt can render real service on emergency.

Aa. 2. Salammoniac brown.

The much keener rust-producing property of salammoniac has long been known. C. W. Sawyer expressly mentions that it was used by the American backwoodsmen for browning their gun barrels. A few trials gave the following results:

a. **Without boiling.** A 0.5% solution produced a satisfactory dark brownish-black after 5 to 6 passes, about equal to that given by common salt *with* boiling. Oiling with vaseline oil and evaporation of the oil at about 150°C. (say 300°F.) recommended.

b. **With boiling.** In reconditioning several rifles from the 1914-18 battlefields, when no other chemical was available, the author obtained a very fine, deep black finish after 3 passes with a 2% salammoniac solution, both on casehardened parts such as rifle bolts and on barrels and furniture: *its appearance was indeed little if at all inferior to that produced by the best "regular" brownes.* No after-rusting provided that the scratching was thorough—and conversely.

This result is quite in accordance with F. Brandeis' statement, that salammoniac will produce as deep a black as can be obtained by any means.

Whether the coating so obtained is as durable as that formed by more sophisticated brownes, the author has not had the opportunity of determining.

Aa. 3. Antimony brown.

This classical browning process was the first to be used on a large scale for army weapons, towards the end of the 18th century: the British musket of that period owed its popular name "Brown Bess" to the colour of its barrel, which was thus treated. The process itself was as follows.

Equal weights of antimony trichloride and a non-drying vegetable oil, e. g. olive or colza, were stirred together, gently warmed, until a perfectly uniform paste resulted, which was applied to the barrels with a woollen rag. After 24 hours they were covered with red rust, when they were oiled and the rust rubbed off. These steps were repeated until a uniform, smooth, sufficiently dark—light to dark reddish-brown—coating without spots or ridges, had been obtained, which took from 10 days to a fortnight, according to the weather.

After the last pass, the barrels were smoothed with a hard-wood polishing stick, scalded (not boiled) with hot water, rubbed with a burnisher, then with linseed oil or wax, and finally varnished with shellac.

According to Hartmann, a black colour is obtained by adding water to the browning paste, scalding, drying rapidly over a charcoal fire, and oiling.

Other sources give different proportions for the mixture, e. g. antimony chloride 1, oil to 3 or 4, chloride 25, oil 4, liq. stibii chlorat. 100, oil 10 parts by weight, the last called "English bronzing salt," which according to Buchner gives a greenish-brown tone, and so on.

This method has long been out of date: according to Brandeis, after-rusting could only be held in check by lacquering, which gradually caused the abandonment of the process.

Aa. 4. Antimony brown. (Brannt & Wahl).

Equal weights of antimony trichloride and ordinary butter are warmed together, and 8 to 10 drops of olive oil added to each 100 grams (abt. 3.5 oz.) of the mixture. After application (with woollen rag) the barrel is warmed over a charcoal fire, soon taking on a fine brown colour. Then rub with buckskin and olive oil, and lacquer with a good amber varnish containing a little shellac.

Aa. 5. Antimony black. (Brannt & Wahl).

Eight parts of antimony trichloride and 4 of sulphuric acid D.1.84 (by weight) are rubbed together with 2 parts of crude wood spirit or gallic acid until uniform, then applied to barrel: scratch when rusted, repeat till colour dark enough, smooth with polishing stick and olive oil and finish by lacquering.

Aa. 6. Antimony blue. (Brannt & Wahl).

Antimony trichloride	2.5 parts by weight.	
Fuming nitric acid	2.5 " " "	
Hydrochloric acid D.1.16	5.0 " " "	

Run the hydrochloric acid drop by drop into the nitric, add the chloride, and after this is completely dissolved apply to parts and rub with green (i. e. freshly cut) oak stick, until the desired colour appears. (No further details of treatment.)

Aa. 7. Instantaneous brown. (Voytier).

Warm barrel until too hot to be held in hand, and rub hard with wash leather on which a few pieces of antimony trichloride have been spread. Repeat till a sufficiently dark colour has been obtained, then finish as described under Aa. 3.

Aa. 8. Zinc browning paste. (Lintner).

Six parts by weight of zinc chloride are rubbed to a uniform paste with 4 parts of olive oil and applied to the warmed barrel. The loose rust is rubbed off with a cloth, repeating application and rubbing until the colour is dark enough. After the last pass rinse with caustic soda lye (5% is strong enough), which wets the surface uniformly, then with soft water and dry. The coating formed is greenish at first, then gradually shading into red and finally brown. After-treatment as for Aa. 2, adding a little "dragon's blood" to the shellac varnish.

Aa. 9. Zinc brown. (Krause).

This is a plain 2% solution of zinc chloride, used in exactly the same way as the ferric chloride brownes B.2 and 3, and giving a brown finish.

Aa. 10. Iodine brown. (Mangeot).

Iodine	15 grs.	0.8 g.
Potassium iodide	37 "	2.0 "
Water t.s. to make	¼ U.S. pt.	100.0 ccm

Ferrous iodide is first formed, which gradually gives place to oxide. No further details in source.

Aa. 11. Twist barrels.

	a.	b.	a.	b.
Copper sulphate	91	110 grs.	5.0	6.0 g.
Spirit of nitre 4%	77	110 "	4.2	6.0 "
Water t.s. to make	¼ U.S. pt.		100.0	100.0 ccm

a. Field & Bonney.   b. Stelle & Harrison.

Working instructions. (b). Let browne stand for a few days after making up. Apply 4 times in each 24 hours, scratching after each application, and repeating until dark enough. A small addition of arsenic (L. acidum arsenicosum, Fr. popularly "mort aux rats," G. "Giftmehl") before the last pass darkens the colour. After last scratching, rub in a mixture of beeswax, boiled linseed oil and turpentine, first with cloth, finishing with the hand.

At room temperature this browne acts very slowly: on soft steel it gave a greyish-black, on hard steel a deeper brownish tone, the parts being steeped in hot logwood decoction (30 grs. per U. S. quart) but not boiled. Arsenic is practically insoluble in the browne.

Aa. 12. Black browne.

	a.	b.	a.	b.
Copper sulphate	110	55 grs.	6.0	3.0 g.
Nitric acid	110	40 "	6.0	2.2 "
Water t.s. to make	¼ U.S. pt.		100.0	100.0 ccm

a. Stelle & Harrison.
b. U. S. Bureau of Standards.

Dissolve salt in about 20 ccm (abt. ¾ fl. oz.) of water, add acid and make up to volume. No further details given.

Aa. 13. Twist barrels.

	a.	b.	a.	b.
Copper sulphate	164	91 grs.	9.0	5.0 g
Nitric acid D.1.42	329	164 "	18.0	9.0 "
Alcohol 90°	219	329 "	12.0	18.0 "
Water t.s. to make	¼ U.S. pt.		100.0	100.0 ccm

a. Buchner, Krause.
b. Michel.

Working instructions. Apply and scratch every 3 to 4 hours, scratching first with fine steel wool, then with hard bristle brush, and repeat till dark enough, then scald with boiling water and oil. According to Michel, (b) gives a black colour with blue shimmer.

Aa. 14. Browne for iron barrels.

	a.	b.	a.	b.
Copper sulphate	146	219 grs.	8.0	12.0 g
Hydrochloric acid	292	438 "	16.0	24.0 "
Spirit of nitre	146	292 "	8.0	16.0 "
Water t.s. to make	¼ U.S. pt.		100.0	100.0 ccm

a. "Metallarbeiter," Vienna.
b. Winckler.

Comparative trial with (a) gave a good but slightly brownish black on soft steel, but of deeper tone on hard steel. No particulars as to concentration of acid given in either source, but (a) made up with acid of 20 G/V-%.

Aa. 15 Twist barrels.

	a.	b.	c.	a.	b.	c.
Copper sulphate	8	14	15 grs.	0.4	0.75	0.8 g
Hydrochloric acid D.1.16	2	3	4 "	0.1	0.15	0.2 "
Nitric acid D.1.42	28	40	55 "	1.5	2.2	3.0 "
Alcohol 90°	55	51	55 "	3.0	2.8	3.0 "
Water t.s. to make		¼ U.S. pt.		100.0	100.0	100.0 ccm

a. Hufschmidt.   b. Krause.   c. Buchner.

After making up, let stand for a few days. All are slightly aqua regia-containing, but mainly nitric, so will not keep active long. According to Buchner (c) is an English browne for twist barrels, but no further instructions in any source.

Aa. 16. Twist barrels. (Winckler).

Copper sulphate	7 grs.	0.4 g
Hydrochloric acid	7 "	0.4 "
Nitric acid	3 "	0.15 "
Spirit of nitre	7 "	0.4 "
Water t.s. to make	¼ U.S. pt.	100.0 ccm

No data as to concentrations or working. Contains aqua regia, so will not keep active for long.

Aa. 17. Organic copper-salt browne. (Mangeot).

Tincture of copper oxalate	15 grs.	0.8 g
Tincture of aloes	9 "	0.5 "
Oxalic acid	18 "	1.0 "
Water t.s. to make	¼ U.S. pt.	100.0 ccm

No further particulars.

Aa. 18. Sublimate browne. ("Shooting & Fishing," about 1898).

This browne, formerly used by the Government Armoury, Springfield, consists of:

Mercuric chloride	55 grs.	3.0 g
Spirit of nitre 4%	2 U.S. fl. oz.	50.0 ccm
Alcohol 95°	2 "	50.0 "

Preparation. First dissolve the chloride in the alcohol, let stand for 6 hours, then add the spirit of nitre. Being practically anhydrous, this browne is frost-proof.

Working instructions. At Springfield the parts to be browned were degreased by rubbing with whiting or powdered gypsum, or by boiling in potash solution: the browne applied with sponge. Duration of rusting 8 hours for the first, 6 hours for each subsequent pass. (Boiling or steaming assumed, but no note on this.) Scratching after each pass with a 9" circular brush of spring-steel wire, 5/1000" (36 Brown & Sharpe gauge), running at 800 revs./min. (peripheral speed 31.4 ft./sec.). Simple oiling after last pass.

Stelle & Harrison give a slightly different composition, and working instructions as follows. Let rust in a warm room for 10 to 12 hours in summer, 15 to 20 hours in winter, rub off rust with cloth and repeat till colour dark enough (no mention of boiling).

At room temperature this browne acts very slowly, therefore best used in connection with (improved or regular) steam chamber, then giving a very good, deep black in 3 to at most 4 passes.

Aa. 19. Black browne. (Hager).

Mercuric chloride	73 grs.	4.0 g
Tartaric acid	6 "	0.3 "
Crude nitric acid, (60-64%)	6 "	0.3 "
Water t.s. to make	¼ U.S. pt.	100.0 ccm

No further particulars.

Aa. 20. Birmingham browne. (twist barrels).

Mercuric chloride	46 grs.	2.5 g
Tincture of galls	23 "	1.25 "
Nitric acid D.1.42	23 "	1.25 "
Water t.s. to make	¼ U.S. pt.	100.0 ccm

Working instructions. Scratch every 2 hours or as soon as rusted, repeating till dark enough: scald with boiling *soft* water (hard water makes the colour fox-red), steep in hot logwood decoction (30 grs. per U. S. quart), dry and oil.

> This browne acts very slowly at room temperature, and after 12 to 15 passes gave an uncertain blackish-brown, but *with* boiling a good acceptable black in 3 to 4 passes.

Aa. 21. Black browne.

	a.	b.	a.	b.
Mercuric chloride	51	137 grs.	2.8	13.0 g
Spirit of nitre (*)	100	474 "	5.5	26.0 "
Alcohol 95°	100	256 "	5.5	14.0 "
Water t.s. to make		¼ U.S. pt.	100.0	100.0 ccm

a. Mangeot. (*). 2%
b. "Arms & Man." (*). 4%

Working instructions (b). Let rust 6 to 12 hours, scratch with fine steel wool, boil with bag of logwood chips, dry and boil.

Aa. 22. Black browne.

	a.	b.	a.	b.
Bismuth oxychloride	37	29 grs.	2.0	1.6 g
Hydrochloric acid. D.1.16	219	164 "	12.0	9.0 "
Alcohol 90°	183	146 "	10.0	8.0 "
Water t.s. to make		¼ U.S. pt.	100.0	100.0 ccm

a. Thirault.
b. Krause.

A particularly excellent browne (given by Krause as for brown colour). Composition (a) is fairly rapid at room temperature. In comparative trial gave a very fine, deep ebony-black on soft and hard steel, as on hardened and case-hardened parts, nearly if not fully equal to that produced by Creusot, Ac.1. Neutralisation needed.

GROUP Ab

Iron-free Brownes with 2 Metallic Salts

Ab. 1. Copper-zinc browne.

	a.	b.	c.		a.	b.	c.
Copper sulphate	46	73	91	grs.	2.5	4.0	5.0 g
Zinc chloride anh.	91	37	46	"	5.0	2.0	2.5 "
Hydrochloric acid D.1.16				A few drops until clear.			
Water t.s. to make	¼ U.S. pt.				100.0	100.0	100.0 ccm

a. Lintner.　　b. Hartmann.　　c. Schützelhofer.

Working instructions. Apply in succession 3 to 4 times, letting dry between each, then scratch when rusted, with soft brush, until the colour is well formed: then after each application. In all from 12 to 15 applications needed. When colour is dark enough, damp with weak caustic soda or potash lye until the surface is thoroughly wet, then boil, preferably in distilled water, smooth with polishing stick or soft iron wire brush, warm to 100°C. (212°F.) and lacquer. The finer finish quite compensates for the higher cost of distilled water.

> Comparative trial showed this browne (composition b) to be somewhat more rapid than Aa. 11, and gave a better final finish. Neutralisation required.

Ab. 2. Sulphate browne. (own composition).

Copper sulphate	110 grs.	6.0 g
Zinc sulphate	73 "	4.0 "
Sulphuric acid D.1.84	91 "	5.0 "
Water t.s. to make	¼ U.S. pt.	100.0 ccm

> Tried by author to compare with the above chloride browne. Gave an acceptable but not specially fine black, and still slightly after-rusting.

Ab. 3. Tin-zinc browne. (own composition).

Stannic chloride anh.	73 grs.	4.0 g
Zinc chloride anh.	73 "	4.0 "
Hydrochloric acid D.1.16	137 "	7.5 "
Spirit of nitre 3%	137 "	7.5 "
Water t.s. to make	¼ U.S. pt.	100.0 ccm

> Comparative trial gave a good, but not exceptionally fine black: better than Ab. 2, and like this, slightly after-rusting.

Ab. 4. Black browne. (Brandeis).

Copper sulphate	18 grs.	1.0 g
Ammonium chloride	73 "	4.0 "
Fuming nitric acid	73 "	4.0 "
Alcohol 90°	55 "	3.0 "
Water t.s. to make	¼ U.S. pt.	100.0 ccm

> Comparative trial gave a very good but faintly brownish-black, which was improved by a 4th pass. Fairly rapid, and very slight tendency to after-rust.

Ab. 5. "English brown." (Voytier).

Antimony trichloride	37 grs.	2.0 g
Mercuric chloride	18 "	1.0 "
Water t.s. to make	¼ U.S. pt.	100.0 ccm

Working instructions. Apply and repeat 10 to 12 times, scratching between each, by which time the parts will have become a chocolate-brown: a reddish tone can be obtained by warming the parts fairly strongly on an iron plate, and a black finish by boiling between each application.

> Hydrochloric acid, just as much as needed, must be added to avoid precipitation of the antimony salt.

Ab. 6. Black browne. (Thirault).

Mercuric chloride	91 grs.	5.0 g
Ammonium chloride	91 "	5.0 "
Water t.s. to make	¼ U.S. pt.	100.0 ccm

> An excellent browne, fairly rapid: comparative trial gave a very fine dull black, comparable to that given by Aa. 22. Neutralisation required.

Ab. 7. Black browne. (Stelle & Harrison).

Mercuric chloride	58 grs.	3.2 g
Copper sulphate	117 "	6.4 "
Spirit of nitre 4%	58 "	3.2 "
Water t.s. to make	¼ U. S. pt.	100.0 ccm

Working instructions. Reapply after standing 1 hour, let stand and scratch after 12 hours, repeat, scratching after each pass, until dark enough, then oil and polish with cloth. (No mention of boiling, but this must be understood: data incomplete).

Ab. 8. Black browne. (Director Roth, Technical School of the Arms Industry, Suhl, Germany).

Mercuric chloride	23 grs.	1.25 g
Zinc nitrate cryst.	22 "	1.2 "
Nitric acid D. 1.42	4 "	0.2 "
Alcohol 90°	55 "	3.0 "
Water t.s. to make	¼ U.S. pt.	100.0 ccm

No further data.

Ab. 9. Black browne. (Voytier).

Mercuric chloride	73 grs.	4.0 g
Copper chloride	37 "	2.0 "
Hydrochloric acid D. 1.16	219 "	12.0 "
Alcohol 90°	183 "	10.0 "
Water t.s. to make	¼ U.S. pt.	100.0 ccm

Working instructions. Apply 2 to 3 times, drying between each, scratch, repeat until dark enough, and when finished, lay for 10 minutes in boiling linseed oil.

According to F. Voytier, this browne gives a very fine, dull black: on account of its similarity to the Creusot browne Ac. 1 (only the bismuth chloride is missing) it may be surmised that this formula was intended to be reproduced, but that a slip has intervened. No mention of boiling after scratching, but obviously needed.

Ab. 10. "20-minute blue."   (J. V. Howe).

Mercuric chloride	146 grs.	8.0 g
Potassium chlorate	82 "	4.5 "
Spirit of nitre 4%	128 "	7.0 "
Water t.s. to make	¼ U.S. pt.	100.0 ccm

After making up, let stand for a few days before using, and after last scratching add a few logwood chips to the boiling water, or lay for ½ hour in hot decoction of logwood. See under Ad. 3 for the general method of operating with these "express" brownes.

Ab. 11. Zinc ammonium browne.   (own composition).

Zinc chloride anh.	137 grs.	7.5 g
Ammonium chloride	91 "	5.0 "
Hydrochloric acid D. 1.16	46 "	2.5 "
Water t.s. to make	¼ U.S. pt.	100.0 ccm

This browne is fairly rapid acting and gives a very fine, deep black on both soft and hard steel with a single rusting. Neutralisation required.

GROUP Ac.

Iron-free Brownes with 3 Metallic Salts

Ac. 1. "Creusot browne."

	a.	b.	c.	a.	b.	c.
Mercuric chloride	37	37	73 grs.	2.0	2.0	4.0 g
Bismuth oxychloride	18	37	37 "	1.0	2.0	2.0 "
Copper chloride	18	37	37 "	1.0	2.0	2.0 "
Hydrochloric acid D. 1.16	110	219	219 "	6.0	12.0	12.0 "
Alcohol 90°	91	183	183 "	5.0	10.0	10.0 "
Water t.s. to make	¼	¼	¼ U. S. pt.	100.0	100.0	100.0 ccm

    a. Michel.
    b. Krause.
    c. "Metallarbeiter", "Deutsche Waffenzeitung".

One of the very best brownes tried out by the author. (See also Aa. 22 and Ab. 6). In comparative trial 3 passes gave a specially fine, deep, durable ebony-black on soft and hard steel, hardened and casehardened parts alike. Fairly rapid at room temperature. Neutralisation required.

The above results were obtained with composition (c). When preparing, remember to add acid before salts, to avoid precipitation of bismuth salt.

Ac. 2. "Nitrate Creusot."

Mercuric nitrate cryst.	73 grs.	4.0 g
Bismuth nitrate neutr. cryst.	128 "	7.0 "
Copper nitrate cryst.	55 "	3.0 "
Nitric acid D. 1.42	183 "	10.0 "
Water t.s. to make	¼ U.S. pt.	100.0 ccm

The special excellence of the Creusot browne suggested the substitution of nitrates for the chlorides, with a view to ascertaining whether this modification would do away with the after-rusting tendency of the chloride browne. A preliminary trial having shown the nitrate browne to be less rapid, the alcohol was omitted.

Comparative trial showed that this modification gave exactly as fine results as the chloride-browne, while after-rusting was effectually eliminated.

Ac. 3. Black browne. ("Metal Industry", N. Y.).

Mercuric chloride	120 grs.	6.6 g
Bismuth oxychloride	40 "	2.2 "
Copper chloride cryst.	40 "	2.2 "
Water t.s. to make	¼ U.S. pt.	100.0 ccm

The composition suggests an incomplete reproduction of the Creusot formula, but the acid, essential to prevent precipitation of the bismuth salt, is omitted.

Ac. 4. Permanganate browne. (own composition).

Mercuric sulphate	9 grs.	0.5 g
Copper sulphate	37 "	2.0 "
Potassium permanganate	7 "	0.4 "
Sulphuric acid D. 1.84	37 "	2.0 "
Water t.s. to make	¼ U.S. pt.	100.0 ccm

Ac. 5. Persulphate browne. (own composition).

Mercuric chloride	9 grs.	0.5 g
Copper sulphate	37 "	2.0 "
Ammonium persulphate	9 "	0.5 "
Water t.s. to make	¼ U.S. pt.	100.0 ccm

These two brownes were tried to ascertain the effect of the two strongly-oxidising salts. In the concentration given both are rather slow-acting, but comparative trial gave a very good, deep black, if not of outstandingly clear superiority. No after-rusting with either. Mercuric chloride necessary in Ac. 5: action would be improved by the addition of a little (say 1.0 g per deciliter of sulphuric acid.

Ac. 6. "Express" browne. (J. V. Howe).

	a	b.	a.	b.
Mercuric chloride	82	91 grs.	4.5	5.0 g
Potassium chlorate	123	91 "	6.75	5.0 "
Potassium nitrate (nitre)	144	91 "	7.9	5.0 "
Spirit of nitre 4%	160	183 "	8.75	10.0 "
Water t.s. to make		¼ U.S. pt.	100.0	100.0 ccm

For working instructions, see Ad. 3. Composition (b) is given as that of a "20-minute browne" and is stated by Mr. Howe to give a satisfactory colour in 5 to 6 passes (see Ad. 3, Remarks, for this claim). When finished, rub with a mixture of equal parts of spermaceti (L. cetaceum album, G. "weisser Amber") and cocoanut oil. H. Lovell, in *American Rifleman*, gives a slightly different composition.

GROUP Ad.

Iron-free Brownes with 4 metallic salts

Ad. 1. Tetrachloride browne. (own composition).

Copper chloride cryst.	55 grs.	3.0 g
Zinc chloride anh.	55 "	3.0 "
Stannic chloride anh.	55 "	3.0 "
Ammonium chloride	37 "	2.0 "
Hydrochloric acid D. 1.16	137 "	7.5 "
Alcohol 90°	137 "	7.5 "
Water t.s. to make	¼ U.S. pt.	100.0 ccm

As brownes Ab. 2 and 3 showed no special points of superiority, this composition was tried to ascertain if the addition of ammonium chloride would better the previous result. Comparative trial gave a very fine, deep black, much superior to that produced by Ab. 2 and 3, and equal to that of browne B. 5 (described under Group B).

In the above concentration this browne is fairly rapid acting: with 4 g per deciliter of each heavy salt it was over-keen, with 2 g per deciliter too dull. No after-rusting.

Ad. 2. "Old English Blue." (American trade "express" browne).

Mercuric chloride	82 grs.	4.5 g
Potassium chlorate	96 "	5.27 "
Potassium nitrate	85 "	4.65 "
Spirit of nitre 4%	See under.	
Water t.s. to make	¼ U.S. pt.	100.0 ccm

This was the composition given by analysis: the quantity of nitrous ether could not be positively determined, different methods giving different results: several trials with quantities varying between 5 and 10 g per deciliter gave no discernably differing result, so that the lower content (if any: see Appendix to Ad. 3) suffices.

The browne moreover contained about 2 g per deciliter of a crumby deposit—intentional or accidental—consisting of a mercurous compound, probably chloride, and potassium chlorate, which did not redissolve on heating.

Working instruction under Ad. 3.

Ad. 3. American "express" browne. (Clyde Baker).

Basic browne.	a.	b.	a.	b.
Mercuric chloride	88	64 grs.	4.8	3.5 g
Potassium chlorate	88	64 "	4.8	3.5 "
Potassium nitrate	44	32 "	2.4	1.75 "
Sodium nitrate	44	32 "	2.4	1.75 "
Spirit of nitre 4%	88	64 "	4.8	2.4 "
Water t.s. to make		¼ U.S. pt.	100.0	100.0 ccm

Composition (a) is for hard crucible steel, (b) for soft steel or iron.

Addenda. For casehardened parts add 0.5 to 1.0 g per deciliter (9 to 18 grs. per ¼ pt.) each of potassium chlorate and nitrate. For barrels of Böhler Antinit, Poldi Anticorro and stainless steel add

Sol. ferric chloride 29%	40 grs.	2.2 g
Nitric acid D. 1.42	44 "	2.4 "
Hydrochloric acid D. 1.16	51 "	2.8 "
Alcohol 90°	55 "	3.0 "

to every deciliter or ¼ pint. From 30 to 40 passes are needed.

Working instructions. These are substantially the same for the "express" brownes and (according to C. Baker) are as follows.

Equipment. A trough of strong black or enamelled sheet (*not* tinned or galvanised sheet) of sufficient size is required; uniformly heated throughout its length by 3 to 4 gas burners or Primus stoves. The trough is filled 2/3 with *soft* water, which is kept *briskly* boiling. The barrels are plugged as usual and handled by the plugs.

A beaker containing 40 to 50 ccm (say 1½ to 2 fl. oz.) is suspended in the trough by means of a suitable hook, so that it dips into the water and takes its temperature.

Pre-pickling. This (according to general rule) is of assistance. After dipping in the bath and taking its temperature, the barrel is lifted out (by two hooks) and wiped over liberally and as rapidly as possible, eventually several times, with a pad of cotton waste or similar material, dipped in the pickle, until its surface has taken on a slightly dull, silvery sheen.

Generally 5% nitric acid is suitable for this purpose, but for Antinit, Anticorro or stainless steel the so-called "Spencer acid" (see Section V.) is to be preferred. If a few bright spots remain, dip again for 5 minutes and go over them with the pad.

Applying browne. A pad made of surgical gauze, held in a split oaken stick (not pine or other resinous wood) and wrapped around it 4 or 5 times, is best suited to applying the browne.

The barrel, heated to boiling temperature, is lifted from the trough and the hot browne applied uniformly and as rapidly as possible to its entire surface, squeezing out any excess of browne on the pad into the beaker. If the barrel was hot enough it dries immediately, that is, in 2 to 3 seconds at most, becoming covered with a bluish-grey coating. Then replace for 2 to 3 minutes in the boiling water.

Scratching. This is done immediately after boiling, barrels preferably with fine steel wool, lightly and uniformly until all loose rust has disappeared, and the light grey colour remains. Hollows, e.g. under a sight base, and others which cannot be reached with steel wool, are scratched with a fine (2/1000″) wire brush.

Repeating. The successive operations of heating, applying browne, reheating and scratching are repeated in order, noting that the *quickest possible* application of the browne is of primary importance. Boiling the coated parts does no harm, so that several pieces can be treated one after the other, and scratched in turn. ("Collective browning", Group G.).

Usually from 3 to 5 passes give an excellent finish, but as many as 12 may be needed for hardened or very thin parts.

After reaching a dark blue-black coating, not appreciably darkening further on repetition, give two more passes, rubbing off with cotton rags instead of scratching (except in hollows, on matted or engraved places), and boil for another 5 minutes.

Finishing. Smooth the browned pieces very gently with a brush of fine iron wire, or if not available, with worn, quite dull steel wool: this heightens the sheen and apparently deepens the tone of the colour. Then apply boiled linseed oil liberally to the still hot pieces, only wiping off when they have quite cooled and the oil has become sticky, then lubricate lightly with gun grease.

The entire procedure, if all goes smoothly, should on the average be completed within about 1½ hours.

Troubles (after C. Baker).

a. If the browne does not "take" in places at the 3d application, degreasing has been faulty. An entirely fresh start must then be made, after pickling off the coating already formed.

b. If the barrel (or other piece) becomes fox-red after the first application, the browne is unsuited to the metal treated. Then try another 1/3 to ½ weaker, or one of the variants given.

c. Incomplete scratching produces—as always—unsightly, sometimes even brown, spots or streaks. Patience.

d. Small parts must be held as close to the surface of the water, and coated as instantaneously as possible, to avoid their rapid cooling. For the same reason the pad of gauze should be left in the hot browne until the very moment of use.

e. Partial or entire disappearance of the coating already formed, when scratching, is usually caused by the parts having become too cold before applying the browne. Three to 4 reapplications, and redipping before each scratching often helps.

d. Unsatisfactory results are *certain* if the coated parts do not dry immediately. Application of browne with all possible speed is a *positive sine quâ non,* to which every attention must be paid.

g. Mr. Baker rightly draws attention to the advantage of pre-etching, which not only assists the browne to "take" but often assists in producing a fine, deep blue-black where an unpleasing reddish-brown would otherwise result (general rule, see under "Rusting", Chapter II. 4).

The author has made a number of trials with both the Baker and the Howe "express" brownes, with test-pieces of soft and hard steel, making the following observations.

1. When made up as prescribed, the salts dissolve only partially, even at boiling temperature, mostly recrystallising out on cooling. The undissolved portion is obviously ineffective in producing colour. On omitting the spirit of nitre they dissolve completely, and in the absence of this ingredient the author obtained a good, deep black in 6 passes, which was visibly better than that obtained in 8 passes with the nitrous ether browne.

2. Whether the addition of ethyl nitrite in presence of such powerful oxidants as potassium chlorate and nitrate is of any real utility, may fairly be questioned, in view of the ease with which it undergoes oxidation, and of its conditional usefulness in brownes (see under "Ethyl nitrite", Chapter IV.).

    It must not be forgotten that "express" brownes contain at most 10% of "spirit of nitre," and that the strongest officinal spirit contains 4.5% of ethyl nitrite, so that the maximum quantity of this ingredient present in a browne is less than 0.5%, and of quite questionable actual effect. Brought to boiling temperature, the alcohol content rapidly distils away, carrying with it the ethyl nitrite, which has a still lower boiling point, so that it disappears first, and any possible effect, it may have had at the beginning quickly vanishes. Given the effect set forth in Note 1, the author concludes that here also the value of ethyl nitrite is at best doubtful, and from the point of solubility of the salts unfavourable. The alcoholic content of the "spirit of nitre" may assist in cold dissolving of the mercuric chloride, but at boiling temperature plain water more than easily suffices.

3. The simultaneous presence of both potassium and sodium nitrates invites question: Mr. Howe states that this ingredient assists in producing a deeper colour (see composition (c) at end of this commentary), a point on which the author has not had sufficient experience to be able to express an opinion.

4. According to the shape and size of the parts to be treated, a complete pass takes from 5 to 8 minutes, so that the designation "20-minute browne" must be understood as "boost" rather than cold fact. Mr. Baker moreover states that he has never been able to get a satisfactory finish within this limit of time.

5. The treatment of larger parts (barrels) and those of com-

plicated shape (e. g. receivers) obviously requires a considerable degree of skill and practice, as it is almost impossible to apply the browne so quickly *and* uniformly as to avoid patchiness, on the first applications at least. (Whether dipping in the hot browne—protecting the surfaces not to be coloured by a suitable varnish—would get over this disadvantage without subsequent trouble, is worth consideration, but its practical inconveniences are obvious). The succeeding passes gradually smooth out this effect—no two passes being exactly alike—but a uniform finish, free from abnormal tone-shading, obviously depends in large measure on the skill of the operator.

6. The undoubted advantage of the process is its rapidity, and it meets a real want in all cases of urgency. Even where this does not apply, as mostly to the private gun owner or enthusiast, this feature no doubt accounts for its apparent popularity, as evidenced by the various trade preparations advertised under fancy names (and mostly at fancy prices).

For the commercial gunmaker on a small or larger scale, it may be questioned, having regard to the high cost of skilled labour in the United States, whether the "express" method of browning is actually cheaper than the classical step-by-step procedure, which, requiring only patience, can be carried out irreproachably by any ordinarily intelligent boy.

Mr. Baker states that he obtains as fine a finish by the "express" as by the classical step-by-step method, and indeed "has no use" for the latter. Every shooter and gunsmith will of course practice the method he prefers, whether on the score of rapidity or quality: it is however noteworthy that according to information kindly afforded the author by an eminent American authority, American luxury gunmakers use the "express" method only when pressed for time, preferring—like their European "opposite numbers"—to adhere to the old step-by-step process in their regular routine.

As to the durability of express-produced coatings, the author has no certain information, or sufficient personal experience of actual use, to be able to give an opinion of any value: while giving it the benefit of the doubt in the meantime, he would consider that its excellence in this respect is, in the words of the proverbial Scot's verdict, "not proven."

Mr. J. V. Howe gives another composition (c) for an "express" browne of this group:

Mercuric chloride	63 grs.	3.45 g
Potassium chlorate	105 "	5.75 "
Potassium nitrate	84 "	4.6 "
Sodium nitrate	84 "	4.6 "
Spirit of nitre	105 "	5.75 "
Water t.s. to make	¼ U. S. pt.	100.0 ccm

This contains a larger amount, relatively, of sodium nitrate (Chili saltpetre), which, Mr. Howe states will produce a deeper colour.

GROUP B

Brownes Containing Iron and 2 Other Metallic Salts
B. 1. Ferric-paste browne (Buchner, Krause).

This is a modification of the old antimony browne, and consists of 10 g of ferric chloride cryst. in 100 ccm (183 grs. in ¼ U. S. pt.) of olive oil ("Eisensalbe" in German). The treatment is the same as for Aa. 3, and the result a reddish-brown coating.

B. 2 and 3. Ferric brownes (E. Beutel).

These consist of 1.5 g ferric chloride cryst. dissolved in 100 ccm (27 grs. in ¼ U. S. pt.) of water, or 2 g of the same salt in 100 ccm (37 grs. in ¼ U. S. pt.) of denatured spirit (free from oily admixture).

Of these three brownes, ferric paste is the slowest-acting, the weak aqueous solution fairly rapid, and the alcoholic midway between. B.2 and 3 give a brown to reddish-brown coating, but the usual black when boiling after each pass.

B. 4. Black browne (Sheffield formula).

Sol. ferric chloride 29%	219 grs.	12.5 g
Alcohol 90°	820 "	45.0 "
Water t.s. to make	¼ U. S. pt.	100.0 ccm

Preparation. Dilute alcohol with a portion of the water, add ferric chloride and make up to volume.

Working instructions. Apply, let parts rust in steam heat 35° to 40°C. (abt. 95-105°F.), boil for 15 to 29 minutes, repeat 2 to 3 times or till dark enough, then oil.

> In comparative trial this very simple browne gave an excellent deep black, albeit rather slowly at room temperature. As it is nothing but a "tincture" of ferric chloride, it has the bad keeping quality of this preparation (see "Tincture of ferric chloride," under "Chemicals," Chapter III.6). This fault is of less importance in continuous trade working, if preparation keeps step, not too far in advance, with requirements. It is very sensibly improved in this respect by the addition of a little nitric or hydrochloric acid (say 1 to 2 g per deciliter=18 to 37 grs. per ¼ U. S. pt.), which also quickens its action, an advantage where no steam chest is available.

B. 5. Dull black browne (Thirault).

Sol. ferric chloride 29%	128 grs.	7.0 g
Nitric acid D. 1.42	8 "	0.4 "
Alcohol 90°	164 "	9.0 "
Water t.s. to make	¼ U. S. pt.	100.0 ccm

Comparative trial gave a very fine deep black. Generally after-rusting, but this tendency very slight on hardened parts.

B. 6. Black browne.

	a.	b.		a.	b.
Sol. ferric chloride 29%	75	120 grs.	4.1	6.6 g
Nitric acid D. 1.42	77	120 "	4.2	6.6 "
Alcohol 90°	86	140 "	4.7	7.7 "
Water t.s. to make		¼ U. S. pt.	100.0	100.0 ccm

    a.   "Instructions for Armourers" (military), 1904 and 1912.
    b.   do.   (naval), 1909.

Browne (a), used at the Royal Small Arms Factory, Enfield, is slow acting at room temperature, more so than British browne C.22 of 1885. On account of acid content better keeping than B.4. and no after-rusting. In comparative trial gave an excellent deep black.

Working instructions for (b) are complicated and not very clear, in particular as regards scratching, and it may be questioned if their final result is appreciably better than that resulting from (a) in comparative trial.

B. 7. A browne which according to Mangeot consists of 2 g each of potassium ferritartrate and tartaric acid in 100 ccm of water (37 grs. in ¼ U. S. pt.).

B. 8. Other organic browne (Mangeot).

Potassium ferritartrate	37 grs.	2.0 g
Oxalic acid	12 "	0.65 "
Acetic acid	73 "	4.0 "
Water t.s. to make	¼ U. S. pt.	100.0 ccm

In comparative trial B.7. gave a brownish, not very pleasing coating after 5 passes. B.8 was almost entirely ineffective. Possibly better with rusting in steam chamber, but not further tried out.

B. 9. Ferrous browne.

Ferrous chloride cryst.	40 grs.	2.2 g
Ether	18 "	1.0 "
Spirit of nitre	27 "	1.5 "
Water t.s. to make	¼ U. S. pt.	100.0 ccm

B. 10 Ferric browne.

Sol. ferric chloride 29%	274 grs.	15.0 g
Spirit of nitre	22 "	1.2 "
Alcohol 95°	310 "	17.0 "
Water t.s. to make	¼ U. S. pt.	100.0 ccm

These two brownes (source uncertain) were not tried out, the absence of acid from the first and the "tincturous" nature of the second not promising favourable results.

B. 11. Twist barrels ("Metallarbeiter").

Ferrous sulphate	91 grs.	5.0 g
Nitric acid D. 1.42	9 "	0.5 "
Ether	9 "	0.5 "
Water t.s. to make	¼ U. S. pt.	100.0 ccm

Known (according to source) as "Page's browne." The (incomplete) working instructions are: apply, scratch with stiff bristle brush when rusted, repeat until colour dark enough, then scald and lacquer, or preferably wax the warmed barrel and rub with woollen cloth until quite cold.

B. 12. Twist barrels   (Krause).

The original formula gives "ferrous nitrate" as a component: in view of the instability of this salt a printer's error may be surmised. On this account, the author tried out two probable compositions, first replacing the ferrous by ferric nitrate, then by its equivalent in ferrous chloride.

a.	Ferrous sulphate (1)	29 grs.	1.6 g
	Sol. ferric nitrate 28%	104 "	5.7 "
	Sol. ferric chloride 29%	164 "	9.0 "
	Water t.s. to make	¼ U. S. pt.	100.0 ccm
	(Slightly oxidised).		
b.	Ferrous sulphate, as above	29 grs.	1.6 g
	Ferrous chloride cryst.	20 "	1.1 "
	Sol. ferric chloride 29%	164 "	9.0 "
	Hydrochloric acid D. 1.16	119 "	6.5 "
	Water t.s. to make	¼ U. S. pt.	100.0 ccm

Hydrochloric acid was added to browne (b) to improve its keeping qualities, and with this addition as shown remained unaltered after 8 years.

In comparative trial browne (a) acted slowly, but on reapplication "took" well, giving a very fine, deep black after only 2 passes (no boiling for twist barrels). No after-rusting.

Browne (b) proved a good deal more rapid than most: periodical inspection needed to avoid spotting and pitting. In comparative trial gave a very good, deep black, equal at least to that given by B.6, in 3 passes. Slight pickling action, so that it somewhat dulls highly polished parts.

B. 13. "Zischang Browne." (J. V. Howe).

This browne, according to Mr. Howe, was used by A. O. Zischang, the famous New York barrel maker, whose products were distinguished by their specially beautiful finish. Specially suited to soft steel.

Preparation. In a 200-ccm round or Erlenmeyer flask 6.4 g (99 grs.) hydrochloric acid D. 1.19, or the equivalent, 7.55 g = 116 grs. of acid D. 1.16, are run, together with 8.0 g (123 grs.) of nitric acid D. 1.42, previously diluted with 25 ccm water, and 1.6 g (abt. 25 grs.) iron filings added in successive small pinches. After the violent reaction (nitrous fumes!) has subsided, the solution is decanted and brought to 100 ccm.

Treatment is the standard one for black browning, boiling and scratching after each pass, and a period of 10 to 14 days must be counted on from beginning to end: the result—for choice on iron or soft steel barrels—is a particularly beautiful, deep brown-black coating, having a kind of translucent shimmer.

In comparative trial the author obtained a very fine black finish on both soft and hard steel.

According to the conventional "aqua regia reaction" (see under "Chemicals, aqua regia"), the approximate composition of this browne would be:

Sol. ferric chloride 29%	183 grs.	10.0 g
Sol. ferric nitrate 28%	164 "	9.0 "
Nitric acid D. 1.42	42 "	2.3 "
Water t.s. to make	¼ U. S. pt.	100.0 ccm

The "salt-prepared" browne is lighter in colour than the "acid-prepared," which is coloured dark brownish-yellow by nitrosyl chloride, and is somewhat keener than the "salt-prepared," but both give equally excellent results.

GROUP C

Brownes containing Iron and 1 other Metallic Salt

C. 1. Antiferric browne (Buchner).

Sol. antimony trichloride (1)	46 grs.	2.5 g
Sol. ferric chloride 29%	46 "	2.5 "
Ether	274 "	15.0 "
Water t.s. to make	¼ U. S. pt.	100.0 ccm

(1). Ph.U.S.A., Brit. or D.Ap.V.4.

C. 2. Antiferrous browne (Beutel).

Antimony trichloride	18 grs.	1.0 g
Ferrous sulphate cryst.	55 "	3.0 "
Ether	18 "	1.0 "
Water t.s. to make	¼ U. S. pt.	100.0 ccm

C. 3. Antiferric browne (Krause).

Sol. antimony trichloride	51 grs	2.8 g
Sol. ferric chloride 29%	128 "	7.0 "
Spirit of nitre 2%	292 "	16.0 "
Water t.s. to make	¼ U. S. pt.	100.0 ccm

C. 4. Antiferric browne (Michel).

Antimony trichloride	914 grs.	50 g
Ferric chloride cryst.	914 "	50 "
Tannin	457 "	25 "
Water t.s. to make	¼ U. S. pt.	100.0 ccm

The mixture should be quite neutral.

C. 5. Antigallic browne.

	a.	b.	c.
Antimony trichloride	2-2.5	4	4-5 parts.
Ferric chloride cryst.	2-2.5	4	4-5 "
Gallic acid	1-1.25	2	2-2.5 "
Water	4-5	1	4 "

    a. Stelle & Harrison.
    b. U. S. Bureau of Standards.
    c. Buchner.

Working instructions.

a. *Stelle & Harrison.* Apply to parts, let dry in warm room, scratch, repeat till coating dark enough, rinse with warm water, dry and rub with linseed oil.

b. According to Fr. Liebetanz the cold mixture, which is a paste rather than a solution, exercises no immediate action, therefore recommends warming the pieces so that they can just be touched by the hand, and letting them rust after application in as warm a room as possible. The parts become a very light blue colour, gradually changing to purple, graphite-grey and finally to a warm brown: scratching after each pass. When dark enough, polish and finish as for Aa.3.

Buchner gives this as a very good, well-tried out browne. According to Hiorns the final colour after many passes is a graphite-grey whereas U. S. Bureau of Standards mentions the blue colour.

According to Krause the pieces should be boiled for 5 to 10 minutes after each alternate pass: colour then black, or nearly so.

    Composition (a) 2-2-1-4 is a dark brown, very turbid liquid, to be thoroughly shaken before use to ensure uniformity. Operating as prescribed by Stelle & Harrison the result was a not specially fine, graphite-grey coating after 5 to 6 passes, which did *not* after-rust.

    At room temperature the parts remain moist for a long time, the high concentration then being liable to cause pitting, from which the scratch brush cannot clear out the rust. Quick drying in heat is therefore needed.

    Boiling and scratching as in the comparative trial, the author obtained an excellent black in 2 to 3 passes.

C. 6. Twist barrels (Krause).

Copper sulphate	9 grs.	0.5 g
Sol. ferric chloride 29%	292 "	16.0 "
Water t.s. to make	¼ U. S. pt.	100.0 ccm

C. 7. Twist barrels (Hartmann).

Copper sulphate	18 grs.	1.0 g
Ferrous sulphate (1)	183 "	10.0 "
Spirit of nitre	37 "	2.0 "
Water t.s. to make	¼ U. S. pt.	100.0 ccm

    (1) oxydised.

C. 8. Twist barrels (Mangeot).

Copper sulphate	9 grs.	0.5 g
Sol. ferric chloride 29%	35 "	1.9 "
Spirit of nitre	110 "	6.0 "
Water t.s. to make	¼ U. S. pt.	100.0 ccm

C. 9. "Ferrioxydin" (German trade browne, from analysis).

Copper sulphate	7 grs.	0.36 g
Sol. ferric chloride 29%	374 "	20.5 "
Nitric acid D. 1.42	268 "	14.7 "
Water t.s. to make	¼ U. S. pt.	100.0 ccm

In comparative trial this gave an excellent deep black on both soft and hard steel.

C. 10. Twist barrels.

General method of operation as described in Section II for twist barrels (brown finish with simple scalding), but black with boiling after each pass. Special instructions below.

	a.	b.	c.	d.	e.	f.
Copper sulphate	6	9	9	9	9	9 grs.
Sol. ferric chloride 29%	23	128	128	137	146	146 "
Nitric acid D.1.42.	111	27	37	37	18	27 "
Alcohol 90°	55	55	55	55	68	55 "
Water t.s. to make	¼	¼	¼	¼	¼	¼ U. S. pt.

Copper sulphate	0.31	0.5	0.5	0.5	0.5	0.5 g
Sol. ferric chloride 29%	1.25	7.0	7.0	7.5	8.0	8.0 "
Nitric acid D.1.42	0.62	1.5	2.0	2.0	1.0	1.5 "
Alcohol 90°	3.0	3.0	3.0	3.0	3.75	3.0 "
Water t.s. to make	100.0	100.0	100.0	100.0	100.0	100.0 ccm

	g.	h.	k.	l.	m.
Copper sulphate	18	37	51	73	110 grs.
Sol. ferric chloride 29%	41	100	150	37	31 "
Nitric acid D.1.42	9	37	51	73	55 "
Alcohol 90°	29	55	59	64	37 "
Water t.s. to make	¼	¼	¼	¼	¼ U. S. pt.

Copper sulphate	1.0	2.0	2.8	4.0	6.0 g
Sol. ferric chloride 29%	2.25	5.5	8.25	2.0	1.7 "
Nitric acid D.1.42	0.5	2.0	2.8	4.0	3.0 "
Alcohol 90°	1.6	3.0	3.25	3.5	2.0 "
Water t.s. to make	100.0	100.0	100.0	100.0	100.0 ccm

a. "American Rifleman".	g. "Instruction sur les Armes Portatives", Brussels 1906.
b. Schuberth.	
c. "Metallarbeiter".	h. Thirault.
d. do.	k. Birmingham black.
e. Thirault.	l. Brannt & Wahl.
f. Hartmann.	m. J. Schön.

To (a). Let parts rust for 3 to 4 hours, scratch with steel wool, repeat 4 times. When dark enough, scald, polish with piece of sacking until cool. Don't leave to rust all night, but scratch and recommence in morning. If figure does not come out well, dilute browne with distilled water and repeat. Add 0.625 g per deciliter (12 grs. in ¼ U. S. pt.) of mercuric chloride in very dry weather. Gives a beautiful brown finish.

(Communicated by Armourer Staff.-sgt. Silver, 1st Royal Dragoons).

To (k). After applying browne, leave parts for

20 minutes in drying oven at 70°C. (160°F.)
20   do.   steam oven.
20   do.   drying oven.

Boil for 20 minutes, scratch, repeat 3 times and fix.

To (1). When rusted, scratch parts firmly with brush dipped in boiling water, repeat twice and lacquer with mixture of

Amber varnish	1 part.
Copal do.	2 do.
Shellac	½ do.
Linseed oil	1 do.

which can be thinned with ½ part turpentine if too thick. After drying, rub with piece of beech charcoal, then with piece of hat felt. Gives a light yellowish-brown finish.

To (m). This is described by J. Schön as "Saxon blue". Let parts dry and repeat till they have taken a fine blue colour, soak for 24 hours in hot water to which a little caustic potash has been added, scratch thoroughly, rinse with the same alkalised water, dry and oil.

C. 11. Twist barrels (same remark as for C. 10).

	a.	b.	c.		a.	b.	c.	
Copper sulphate	51	383	510	grs.	2.8	21.0	28.0	g
Sol. ferric nitrate 28%	27	210	270	"	1.5	11.5	15.0	"
Nitric acid D.1.42	0.6	35	47	"	0.03	1.9	2.6	"
Alcohol 90°	26	210	255	"	1.4	11.5	14.0	"
Water t. s. to make	¼	¼	¼	U. S. pt.	100.0	100.0	100.0	ccm

a.   Hufschmidt.
b.   U. S. Bureau of Standards.
c.   Buchner.

Working instructions (b. & c.) Let rust for 24 hours, scratch and repeat till dark enough. Dip in boiling water slightly alkalised with caustic soda, to neutralise all acid, rub lightly with polishing stick, warm to 100°C. (212°F.) and lacquer with spirit shellac varnish to which a little dragon's blood is added, and smooth with burnisher after cooling.

C. 12. Twist barrels   (Stelle & Harrison).

Copper sulphate	18 grs.	1.0 g
Sol. ferric chloride 29%	44 "	2.4 "
Spirit of nitre 4%	106 "	5.8 "
Alcohol 90°	58 "	3.1 "
Water t.s. to make	¼ U. S. pt.	100.0 ccm

Action can be accelerated by the addition of 0.025 to 0.04 g mercuric chloride per deciliter (½ to ¾ gr. per ¼ U. S. pt.). Add a little milk of lime to scalding water, to neutralise all acid.

C. 13. Browne of Swiss Federal Armoury, *Bern*, (from analysis).

Copper sulphate	33 grs.	1.8 g
Sol. ferric chloride 29%	227 "	12.5 "
Nitric acid D.1.42.	132 "	7.2 "
Spirit of nitre	See below.	
Water t.s. to make	¼ U. S. pt.	100.0 ccm

A specially excellent browne, fairly rapid at room temperature, giving in comparative trial a very fine deep black on soft or hard steel, as on hardened and casehardened parts.

The freshly-prepared browne is a clear, grass-green liquid, slightly smelling, which gradually darkens by the reaction of nitric acid on the ethyl nitrite, the nitrous oxides evolved first shading its colour to olive-green, and finally to a greenish-brown. Exposure to direct sunlight promotes this action, which is then ended in about six months (and about 2½ years in diffused light), its termination being distinguished by the disappearance of the nitrous smell.

At the beginning of and during this reaction the browne is fairly rapid acting, duller after its termination, but as excellent in point of final result. After the first application and drying, a reapplication after about 2 hours is recommended to activate action.

There is no after-rusting when scratching has been thorough: the browne has good keeping qualities, and only deposits slightly after standing for about 2½ years.

On account of the internal reaction, the composition keeps changing, and the ethyl nitrite content varies with the age of the sample analysed, so that its determination gives diverging results. After a series of trials with different proportions, the author found that the addition of 2.5 g per deciliter (46 grs. per ¼ U. S. pt.) of 15% ethyl nitrite (I.. spiritus aetheris nitrosi 15%) most closely approximated in appearance and result to a sample fresh from the Armoury.

The author also tried variants of this excellent browne, slightly modifying the composition. Substituting alcohol (2.5 g per deciliter = 46 grs. per ¼ U. S. pt.) for the ethyl nitrite, the browne was slower in action, but otherwise equally good. Replacing the pure by its equivalent of *fuming* nitric acid (5.3 g per deciliter = 97 grs. per ¼ U. S. pt.) and omitting alcohol and ethyl nitrite, rendered it keener: it then keeps permanently and according to the nature of metal, often gives a splendid black with only 2 passes.

C. 14. Twist barrels (same remark as for C. 10).

	a.	b.	c.	d.	e.	f.
Copper sulphate	7	20	26	33	37	37 grs.
Sol. ferric chloride 29%	2	155	40	49	73	73 "
Nitric acid D.1.42	29	78	49	23	26	55 "
Spirit of nitre	29	237	73	33	37	55 "
Alcohol 90°	7	220	128	66	73	275 "
Water t.s. to make	¼	¼	¼	¼	¼	¼ U.S.pt.

	a.	b.	c.	d.	e.	f.
Copper sulphate	0.4	1.1	1.4	1.8	2.0	2.0 g
Sol. ferric chloride 29%	0.1	8.5	2.2	2.7	4.0	4.0 "
Nitric acid D.1.42	1.6	4.25	2.7	1.25	1.4	3.0 "
Spirit of nitre	1.6	13.0	4.0	1.8	2.0	3.0 "
Alcohol 90°	0.4	12.0	7.0	3.6	4.0	15.0 "
Water t.s. to make	100.0	100.0	100.0	100.0	100.0	100.0 ccm

	g.	h.	k.	l.	m.	n.
Copper sulphate	46	49	82	84	91	91 grs.
Sol. ferric chloride 29%	100	128	91	88	23	23 "
Nitric acid D.1.42	46	26	27	155	11	23 "
Spirit of nitre	46	26	27	210	11	23 "
Alcohol 90°	200	49	55	41	11	46 "
Water t.s. to make	¼	¼	¼	¼	¼	¼ U.S.pt.

	g.	h.	k.	l.	m.	n.
Copper sulphate	2.5	2.7	4.5	4.6	5.0	5.0 g
Sol. ferric chloride 29%	5.5	7.0	5.0	4.8	1.25	1.25 "
Nitric acid D.1.42	2.5	1.4	1.5	8.5	0.6	1.25 "
Spirit of nitre	2.5	1.4	1.5	11.5	0.6	1.25 "
Alcohol 90°	11.0	2.7	3.0	2.25	0.6	2.5 "
Water t.s. to make	100.0	100.0	100.0	100.0	100.0	100.0 ccm

	o.	p.	q.	r.	s.	t.
Copper sulphate	91	102	110	110	115	274 grs.
Sol. ferric chloride 29%	79	97	55	183	24	137 "
Nitric acid D.1.42	33	26	27	274	27	64 "
Spirit of nitre	9	26	27	274	27	137 "
Alcohol 90°	18	26	64	55	91	137 "
Water t.s. to make	¼	¼	¼	¼	¼	¼ U.S.pt.

	o.	p.	q.	r.	s.	t.
Copper sulphate	5.0	5.6	6.0	6.0	6.3	15.0 g
Sol. ferric chloride 29%	4.3	5.3	3.0	10.0	1.3	7.5 "
Nitric acid D.1.42	1.8	1.4	1.5	15.0	1.5	3.5 "
Spirit of nitre	0.5	1.4	1.5	15.0	1.5	7.5 "
Alcohol 90°	1.0	1.4	3.5	3.0	5.0	7.5 "
Water t.s. to make	100.0	100.0	100.0	100.0	100.0	100.0 ccm

a. "Arms & the Man."	l.	Krause.
b. Birmingham brown.	m.	J. Schön (Dupin).
c. Stelle & Harrison.	n.	Hiorns.
d. Krause.	o.	Field & Bonney.
e. Hartmann.	p.	Mangeot.
f. "Metallarbeiter."	q.	"Golding's blue-black."
g. Mangeot.	r.	Hartmann.
h. do.	s.	Stelle & Harrison.
k. Lintner.	t.	C. W. Sawyer.

To (a). According to some slow but excellent.

To (b). "Military black": operation as for C.10.k.

To (m). This browne, cited by J. Schön (without statement of water quantity), is from the work of Naval Officer C. Dupin, Member of the Institute of France, on the British forces (English translation 1822), who gives the following working instructions.

Apply and let parts rust for 24 hours, scratch thoroughly with hard (bristle) brush, repeat 2 to 3 times, until the barrel has acquired a good brown. Dip in boiling water, slightly alkalinised to neutralise all acid, smooth with hardwood polishing stick, warm to 100°C. (212°F.) and lacquer with varnish of

Shellac	46 grs.	2.5 g
Dragon's blood	9 "	0.5 "
Alcohol 90°	¼ U.S.pt.	100.0 ccm

and after drying smooth with burnisher.

To (q). In comparative trial this browne did produce a markedly bluish-black: slow-acting but giving excellent final result.

C. 15. Twist barrels (Krause).

Copper sulphate	73 grs.	4.0 g
Sol. ferric chloride 29%	77 "	4.2 "
Nitric acid D.1.42	18 "	1.0 "
Spirit of nitre 2%	23 "	1.25 "
Ether	49 "	2.7 "
Alcohol 90°	49 "	2.7 "
Water t.s. to make	¼ U.S.pt.	100.0 ccm

C. 16. Twist barrels (Stelle & Harrison).

Copper sulphate	44 grs.	2.4 g
Sol. ferric chloride 29%	20 "	1.1 "
Tincture of benzoin	44 "	2.4 "
Spirit of nitre 4%	183 "	10.0 "
Alcohol 90°	183 "	10.0 "
Water t.s. to make	¼ U.S.pt.	100.0 ccm

Treatment as for J. Schön (Dupin) C.14.m.

C. 17. Swiss black (Beutel).

Copper sulphate	22 grs.	1.2 g
Ferrous sulphate cryst.	55 "	3.0 "
Sol. ferric chloride 29%	55 "	3.0 "
Alcohol 90°	91 "	5.0 "
Water t.s. to make	¼ U.S.pt.	100.0 ccm

Regular method for black colour. This browne is qualitatively the same as "Ferrobronze" C.19, adding alcohol and omitting hydrochloric acid, neither modification enhancing its keeping qualities. According to Herr Beutel 1.2 g mercuric chloride in 100 ccm (22 grs. in ¼ U. S. pt.) can be substituted for the ferric chloride.

C. 18. Twist barrels (Krause).

Copper nitrate cryst.	22 grs.	1.2 g
Ferrous sulphate cryst.	55 "	3.0 "
Sol. ferric chloride 29%	57 "	3.1 "
Water t.s. to make	¼ U.S.pt.	100.0 ccm

Working instructions. Apply every 12 hours, boil after each alternate application, scratch and repeat till colour dark enough.

After 6 passes, summary scalding.

If boiled after each alternate pass, according to Herr Krause's general instruction for brown colouring, a black finish results. The above treatment is stated to give a very agreeable, reddish-brown shade.

The browne deposits slightly after standing for about a year, as it contains no acid.

C. 19. "Ferrobronze" (French trade browne, from analysis).

Copper sulphate	22 grs.	1.2 g
Ferrous sulphate cryst.	73 "	4.0 "
Sol. ferric chloride 29%	102 "	5.6 "
Hydrochloric acid D.1.16	9 "	0.5 "
Water t.s. to make	¼ U.S.pt.	100.0 ccm

Working instructions. Boil parts for 20 minutes when the loose rust has formed, scratch, repeat 3 to 4 times, and oil after last scratching.

This is a specially excellent browne, producing a very fine, deep black on soft and hard steel, hardened and casehardened parts, and free from after-rusting.

The browne is a clear, grass-green liquid of good keeping quality, but which deposits slightly after long standing. The addition of hydrochloric acid, up to 5 g per deciliter (91 grs. per ¼ U. S. pt.) renders it permanently keeping and quicker in action, but inclining to after-rusting in about the proportion of this additional acid.

C. 20. "Swiss black" ("Metallarbeiter").

Copper sulphate	7 grs.	0.4 g
Sol. ferric chloride 29%	420 "	23.0 "
Nitric acid D.1.42	26 "	1.4 "
Hydrochloric acid D.1.16	7 "	0.4 "
Water t.s. to make	¼ U.S.pt.	100.0 ccm

Working instructions. After application, let parts stand for
20 minutes in drying oven at 40°C. (105°F.)
30   "   " steam chamber at 40°C.
20   "   " drying oven   "   "

then boil for 20 minutes, scratch and repeat 3 to 4 times, when
they will have acquired a fine, deep black, like that of the
steel cases of Swiss watches.

C. 21. Black browne (Krause).

Copper sulphate	46 grs.	2.5 g
Sol. ferric chloride 29%	457 "	25.0 "
Nitric acid D.1.42	91 "	5.0 "
Hydrochloric acid D.1.16	91 "	5.0 "
Alcohol 90°	82 "	4.5 "
Water t.s. to make	¼ U.S.pt.	100.0 ccm

This browne is qualitatively the same as C.20 with added alcohol,
therefore somewhat slower in action. In comparative trial gave
a very fine, dull black.

Note. Brownes C.20 and 21 contain aqua regia, so that
their composition is continually altering, as described under
this heading "Chemicals," Chapter III. Prepare in small quantity, and use at a definite, uniform age to make sure of comparable results.

C. 22. Browne of Royal Small Arms Factory, Enfield, 1885.

Copper sulphate	46 grs.	2.5 g
Sol. ferric chloride 29%	47 "	2.6 "
Nitric acid D.1.42	46 "	2.5 "
Sulphuric acid D.1.84	35 "	1.9 "
Spirit of nitre 2.5%	91 "	5.0 "
Alcohol 90°	91 "	5.0 "
Water t.s. to make	¼ U.S.pt.	100.0 ccm

Working instructions. Let parts rust for 1½ hours in steam
oven, boil, scratch, repeat 4 times and oil.

In comparative trial gave a very good, deep black: quicker
in action than the newer British browne B.6. and free from
after-rusting.

This browne differs qualitatively from C.14 only by the
addition of sulphuric acid.

C. 23. Brown colour (Mangeot).

Mercuric nitrate cryst.	73 grs.	4.0 g
Ferriacetate tincture	73 "	4.0 "
Nitric acid D.1.42	t.s. to dissolve.	
Water t.s. to make	¼ U.S.pt.	100.0 ccm

C. 24. Black browne ("Arms & the Man").

	a.	b.	a.	b.
Mercuric chloride	4	49 grs.	0.2	2.7 g
Sol. ferric chloride 29%	2	22 "	0.15	1.2 "
Spirit of nitre 4%	6	49 "	0.3	2.7 "
Alcohol 90°	4	110 "	0.2	6.0 "
Water t.s. to make		¼ U.S.pt.	100.0	100.0 ccm

Working instructions. Composition (a) is for twist, (b) for steel barrels. Let parts rust for 6 to 12 hours, scratch with steel wool, repeat from 6 to 12 times, dip in boiling logwood decoction, dry and oil.

C. 25. Black browne (Schützelhofer).

Mercuric chloride	55 grs.	3.0 g
Sol. ferric chloride 29%	457 "	25.0 "
Nitric acid	274 "	15.0 "
Alcohol 90°	365 "	20.0 "
Water t.s. to make	¼ U.S.pt.	100.0 ccm

Working instructions. Keep parts in steam cupboard (not under 35°C., 95°F.) for at least 1 hour, boil for 20 minutes, scratch and repeat till dark enough. After last scratching lay in hot water with added logwood decoction, which must *not boil,* dry and oil.

  Note. No concentration of acid is given, but the mercuric chloride in the impossible amount of 30 g per deciliter, so that the actual composition of the browne is a matter of conjecture.

C. 26. Birmingham browne for twist barrels.

Mercuric chloride	23 grs.	1.25 g
Sol. ferric chloride 29%	23 "	1.25 "
Nitric acid D.1.42	6 "	0.3 "
Spirit of nitre 2.5%	46 "	2.5 "
Alcohol 90°	9 "	0.5 "
Water t.s. to make	¼ U.S.pt.	100.0 ccm

Working instructions. Let parts rust overnight, scratch, reapply twice at intervals of 3 to 4 hours and let rust overnight. Repeat this treatment.

Apply again twice at 3 to 4-hour intervals, and after the second scratching lay for ½ minute in boiling decoction of logwood (2 g per liter = 30 grs. per U. S. quart), to which 0.5 g (7½ grs.) of ferrous sulphate have been added.

Wipe dry, apply browne to the still hot barrels and let rust overnight.

Scratch, apply browne cold, let stand ½ hour, dip quickly twice in boiling water, scratch again. Repeat hot application, rusting and scratching, finally with browne diluted to half its strength, until the desired colour is obtained.

Own trial. This brown is specially adapted to twist barrels, is slow acting, and after 5 to 8 passes (no boiling) gave an excellent uniform brown colour, which shades into a warm, agreeable reddish plum-brown after the logwood treatment. If boiled for a longer time the usual black colour results.

The figure is faintly distinguishable, and must be further brought out by one or other of the methods detailed in Chapters II and V under the heading "Twist barrels."

When carefully carried through, this browne, the use of which visibly demands considerable patience, gives a correspondingly fine result on twist barrels.

C. 27. Black browne ("Metallarbeiter").

Mercuric chloride	23 grs.	1.25	g
Sol. ferric chloride 29%	55 "	3.0	"
Nitric acid D.1.42	23 "	1.25	"
Spirit of nitre 2.5%	91 "	5.0	"
Alcohol 90°	91 "	5.0	"
Water t.s. to make	¼ U.S.pt.	100.0	ccm

In comparative trial this browne gave a very good black, the same result being obtained when the spirit of nitre was replaced by ethyl *nitrate* (not further affected by nitric acid) or plain alcohol.

C. 28. Black browne (Dr. Wogrinz, Vienna).

Mercuric chloride	37 grs.	2.0	g
Ferrous chloride cryst.	128 "	7.0	"
Sol. ferric chloride 29%	38 "	2.1	"
Hydrochloric acid D.1.16	A few drops.		
Water t.s. to make	¼ U.S.pt.	100.0	ccm

Working instructions. After applying browne, let parts stand in drying oven for 20 to 30 minutes, then in steam of boiling water for the same time, after which boil for 20 minutes. Scratch with fine iron wire (2½ 1000″) brush, repeat till dark enough, and oil with hot linseed oil. No after-rusting.

In comparative trial (rusting at room temperature) this browne gave a very fine, deep black: for less important pieces 2 passes are sufficient to give a very acceptable colour.

Dr. Wogrinz states that a browne should contain both ferrous and ferric salts (see also C.19) and these two examples, although of course not exclusive, speak well for this contention.

C. 29. Black browne (J. V. Howe).

Mercuric chloride	23 grs.	1.25	g
Ferrous sulphate cryst.	18 "	1.0	"
Sol. ferric chloride 29%	28 "	1.55	"
Crude sulphur	23 "	1.25	"
Nitric acid D.1.42	15 "	0.8	"
Sulphuric acid D.1.84	69 "	3.8	"
Spirit of nitre 4%	54 "	2.9	"
Alcohol	37 "	2.0	"
Water t.s. to make	¼ U.S.pt.	100.0	ccm

This browne is qualitatively identical with D.18, omitting copper sulphate and adding sulphuric acid: ordinary trade concentration of acids assumed. For effect of crude sulphur addition, see under D.18, this browne requiring the usual boiling of rusted parts to obtain black finish.

GROUP D

Brownes containing Iron and 2 other metallic Salts

D. 1. Brown colour ("Metallarbeiter").

Sol. antimony trichloride D.1.35	22 grs.	1.2 g
Sol. copper chloride (1)	33 "	1.8 "
Sol. ferric chloride 29%	55 "	3.0 "
Spirit of nitre 2.5%	55 "	3.0 "
Water t.s. to make	¼ U.S. pt.	100.0 ccm

Concentration of (1) not stated. Recommended for twist barrels, but no further details given.

D. 2. Black browne (Beutel).

Antimony trichloride	15 grs.	0.8 g
Copper chloride cryst.	29 "	1.6 "
Sol. ferric chloride 29%	457 "	25.0 "
Nitric acid D.1.42	15 "	0.8 "
Water t.s. to make	¼ U.S. pt.	100.0 ccm

Working instructions. Apply and scratch several times in succession, until the brown colour is dark enough: then boil or lay pieces on wire netting in the vapour of boiling water, so as to expose them to the steam. This is stated to prevent spotting more effectively than boiling. Repeat as necessary, then oil.

D. 3. Black browne

	a.	b.	a.	b.
Sol. antimony trichloride D.1.35	29	128 grs.	1.6	7.0 g
Copper sulphate	73	40 "	4.0	2.2 "
Sol. ferric chloride 29%	22	8 "	1.2	0.45 "
Nitric acid D.1.42	73	20 "	4.0	1.1 "
Alcohol 90°	37	24 "	2.0	1.3 "
Water t.s. to make	¼ U.S. pt.		100.0	100.0 ccm

a. Hartmann.   b. Hager.

D. 4. Brown colour (Brannt & Wahl).

Antimony trichloride	40 grs.	2.2 g
Copper sulphate	55 "	3.0 "
Sol. ferrous sulphate (1)	40 "	2.2 "
Spirit of nitre 2.5%	55 "	3.0 "
Water t.s. to make	¼ U.S. pt.	100.0 ccm

Concentration of (1) not stated. On account of the small absolute amount of salt, a saturated solution at 15°C. can be assumed, i. e. about 1.5 g cryst. ferrous chloride in 100 ccm (27 grs. in ¼ U. S. pt.).

Working instructions. Let parts rust for 24 hours, scratch and repeat twice. After last scratching, polish with wash leather and olive oil, let dry for 24 hours and repeat oil polish.

D. 5. German Government Armouries' black browne (Hager).

Copper sulphate	91 grs.	5.0 g
Cream of tartar (1)	60 "	3.3 "
Sol. ferric chloride 29%	31 "	1.7 "
Nitric acid D.1.42	31 "	1.7 "
Spirit of nitre 2%	31 "	1.7 "
Water t.s. to make	¼ U.S. pt.	100.0 ccm

No further details given.   (1) Potassium bitartrate.

D. 6. Black browne (R. Zitte).

Mercuric chloride	27 grs.	1.5 g
Copper sulphate	27 "	1.5 "
Ferrous chloride cryst.	37 "	2.0 "
Alcohol 90°	102 "	5.6 "
Water t.s. to make	¼ U.S. pt.	100.0 ccm

Working instructions. Let parts rust in warm room (25° to 35° C. = 75° to 95°F.), scratch with soft wire brush, and repeat every 4 hours (4 times daily). Boil for 15 minutes after each 4th rusting.

These operations are repeated for 5 or 6 days, until the desired colour is obtained, the parts then rubbed with vaseline. The coating resulting is very durable, fine black and occasionally blue-black.

Comparative trial gave a good, deep black on soft and hard steel with 3 to 4 passes.

Remarks. Herr Zitte states that the browne is "extraordinarily stable", but the author noted the beginning of deposit within a few days after preparation, which is not surprising considering the easy oxidisability of ferrous salt and the absence of free acid: after standing 6 months this had increased to an amount such that acidifying with 2 g hydrochloric acid D. 1.16 (or 3.5 g sulphuric acid D. 1.84) per deciliter (37 and 64 grs. respectively per ¼ U.S. pt.) was needed to redissolve. The acidified browne then remained unaltered for 3 years.

When freshly prepared the browne is greenish-blue, but with advancing oxidation gradually changes to pale yellow.

D. 7. Black browne.

	a.	b.	a.	b.
Mercuric chloride	9	27 grs.	0.5	1.5 g
Copper sulphate	5	91 "	0.25	5.0 "
Sol. ferric chloride 29%	102	547 "	5.7	30.0 "
Alcohol 90°	27	547 "	1.5	30.0 "
Water t.s. to make		¼ U.S. pt.	100.0	100.0 ccm

a. Mangeot.   b. Schützelhofer.

Note. Composition (b) quoted by Schützelhofer from the *Schweizerische Werkmeister-zeitung*, can only be considered as possible (his formulae are not distinguished for clearness), as he gives the improbable amount of 15 g mercuric chloride per deciliter (274 grs. per ¼ U.S. pt.). With its large ferric chloride and alcohol ("tincture") content such a browne has not good keeping qualities. No further details given.

D. 8. Twist barrels.   (see note to C. 10).

	a.	b.	c.	d.
Mercuric chloride	1	4	48	82 grs.
Copper sulphate	18	24	48	110 "
Sol. ferric chloride 29%	44	20	33	46 "
Spirit of nitre	100	91	95	91 "
Alcohol 90°	183	119	46	18 "
Water t.s. to make	¼	¼	¼	¼ U.S. pt.

Mercuric chloride	0.05	0.2	2.6	4.5 g
Copper sulphate	1.0	1.3	2.5	6.0 "
Sol. ferric chloride 29%	2.4	1.1	1.8	2.5 "
Spirit of nitre	5.5	5.0	5.2	5.0 "
Alcohol 90°	10.0	6.5	2.5	1.0 "
Water t.s. to make	100.0	100.0	100.0	100.0 ccm

a. C. W. Sawyer.
b. *Arms & the Man.*
c. C. W. Sawyer.
d. Brannt & Wahl.

To (b). Rust for 6 to 12 hours, scratch with finest wire brush, repeat 6 to 12 times, scald with boiling logwood decoction, dry and oil.

To (d). Rust for 24 hours, scratch, brush with hard bristle brush and hot water and repeat twice more. After last scratching, polish with wash leather and olive oil to which a little turpentine has been added, let dry for 12 hours and finally polish with olive oil.

D. 9. Twist barrels. (see note to C. 10).

	a.	b.	c.	d.	e.	f.
Mercuric chloride	9	9	11	11	15	15 grs.
Copper sulphate	6	18	6	5	7	7 "
Sol. ferric chloride 29%	5	164	16-22	44	24	24 "
Nitric acid D. 1.42	9	9	11	11	15-26	15 "
Alcohol 90°	2	3	51	44	128	31 "
Water t.s. to make	¼	¼	¼	¼	¼	¼ U.S.pt.

Mercuric chloride	0.5	0.5	0.6	0.6	0.8	0.8 g
Copper sulphate	0.3	1.0	0.3	0.25	0.4	0.4 "
Sol. ferric chloride 29%	0.25	9.0	0.9-1.2	2.4	1.3	1.3 "
Nitric acid D. 1.42	0.5	0.5	0.6	0.6	0.8-1.4	0.8 "
Alcohol 90°	0.1	1.5	2.8	2.4	40.0	1.7 "
Water t.s. to make	100.0	100.0	100.0	100.0	100.0	100.0 ccm

	g.	h.	k.	l.	m.	
Mercuric chloride	18	26	33	46	55	grs.
Copper sulphate	9	13	16	23	33	"
Sol. ferric chloride 29%	33	41	23	37-46	219	"
Nitric acid D. 1.42	18-27	26-38	33	46	47	"
Alcohol 90°	97	155	55	192	183	"
Water t.s. to make	¼	¼	¼	¼	¼	U.S.pt

	g.	h.	k.	l.	m.	
Mercuric chloride	1.0	1.4	1.8	2.5	3.0	g
Copper sulphate	0.5	0.7	0.9	1.25	1.8	"
Sol. ferric chloride 29%	1.8	2.25	1.25	2.0-2.5	12.0	"
Nitric acid D. 1.42	1.0-1.5	1.4-2.1	1.8	2.5	2.6	"
Alcohol 90°	5.2	8.5	3.0	10.5	10.0	"
Water t.s. to make	100.0	100.0	100.0	100.0	100.0	ccm

a. F. Voytier, black browne, so-called "English colour".
b. Liége black browne.
c. Greener 1846 and 1910, twist barrels.
d. Liége black browne.
e. Stelle & Harrison, twist barrels.
f. Stelle & Harrison, twist barrels.
g. "Metallarbeiter".
h. Stelle & Harrison, twist barrels.
k. "Bazaar", London.
l. Greener, black browne for steel barrels.
m. Hager.

To (b). Black browne for hard crucible and special steel, kindly communicated by MM. Auguste Francotte, Liége. Standard mode of operation (as comparative trial). 4 passes give an excellent deep black finish on Poldi-Anticorro.

To (c). Rusting in steam chamber, 18 hours for the first, 10 to 12 hours for subsequent passes, scalding between passes with slight addition of caustic soda, copper sulphate or logwood decoction. Hard barrels with much steel require a weaker browne and longer rusting periods. Repeat until a deep brown colour has been obtained.

With slow rusting and addition of 0.62 g per deciliter (11 grs. per ¼ U.S. pt.) of crude sulphur (also called "black brimstone", because really black), a fine plum-brown colour results.

According to J. H. Walsh ("Stonehenge" of the London "Field") apply browne every 3 to 4 hours and scratch every morning and evening: repeat until dark enough, scald and polish with wash leather.

To (d). MM. Auguste Francotte's black browne for usual trade qualities of steel; standard operation for black finish, 3 passes being usually sufficient for a very good result.

To (e). Let browne stand for 6 weeks after preparation before using.

To (f). Same remark. Apply every 2 hours and scratch every morning.

To (h). Same as (e) if quantities in apothecaries' weight: not clear.

To (m). Source gives 3.0 g per deciliter (55 grs. per ¼ U.S. pt.) of *crude* nitric acid (abt. 61%): see under this heading, Chapter III. "Chemicals", for difference in effectiveness compared to pure nitric acid.

D. 10. Springfield Armoury browne.   (Colvin & Viall).

Apart from a more or less accidental intrusion, this browne belongs qualitatively to the next group, but will be treated here on account of its unusual mode of preparation.

Preliminary. According to the original source, the first step is the preparation of a "tincture of steel", for which purpose 3 lbs. of "carbonate of iron" are put into a stoneware jar and 3 quarts (U.S.) of hydrochloric acid (assumed D. 1.19) added. After the acid has dissolved all the "carbonate" it can take up, the clear liquid is decanted and 9 quarts of grain alcohol added.

The browne then consists of:

Mercuric chloride	1	ounce
Copper sulphate	½	"
"Tincture"	6	"
Nitric acid D. 1.42	1	"
Spirit of nitre 4%	6	"
Water	2	quarts.

Reducing these quantities to proportions more manageable for the private shooter and gunsmith, 20 g of "carbonate of iron" correspond to 41.67 ccm of hydrochloric acid D. 1.19 and 125 ccm of alcohol 95°.

> *Author's reproduction.* The so-called "carbonate of iron" (according to the statements of several chemical firms * and his own determinations) consists mainly of ferric oxide and hydroxide, with quite insignificant, and consequently negligible, traces of ferrous compound.
>
> Hydrochloric acid D. 1.19 not being available, its equivalent of acid D. 1.16 was taken, 50 ccm of which, added to 120 ccm of absolute (99.5%) alcohol instead of 95° spirit, reestablish within a very small fraction the quantity proportions of the original prescription.
>
> After completion of the reaction an average of 40 ccm of liquid D. 1.37 (ferric chloride content abt. 36 G-%) was decanted: according to the physical state of the "carbonate", a small, variable residue of hydrochloric acid remains free, so that the average composition of the "tincture" (in round numbers) can be taken as:
>
> | | | |
> |---|---|---|
> | Ferric chloride anh. | 12.5 | G-% |
> | Hydrochloric acid D. 1.16 | 2.5 | " |
> | Alcohol, absolute | 63.0 | " |
> | Water | 22.0 | " |

---

* This was done in England.

Slight variations in composition, due to its origin and mode of preparation, are moreover without the least practical importance in view of the small quantities utilised.

The browne then consists of:

Mercuric chloride	1.26 g
Copper sulphate	0.63 "
"Tincture"	7.5 "
Nitric acid D. 142	1.26 "
Spirit of nitre 4%	7.5 "
Water t.s. to make	100.0 ccm

or finally:

Mercuric chloride	23 grs.	1.26 g	
Copper sulphate	11.5 "	0.63 "	
Sol. ferric chloride 29%	58 "	3.2 "	
Hydrochloric acid D. 1.16 abt.	3.5 "	0.19 "	
Nitric acid D. 1.42	23 "	1.26 "	
Spirit of nitre 4%	137 "	7.5 "	
Alcohol 90°	100 "	5.5 "	
Water t.s. to make	¼ U.S. pt.	100.0 ccm	

Working instructions. Three passes of 1 hour each in steam chamber, boiling for 5 minutes after each, scratching and final oiling, are sufficient to obtain a good, deep colour.

Note 1. When used as prescribed, and also in comparative trial, the browne gave a very good, deep black on soft and hard steel. At room temperature it acts slowly and 2 to 3 re-applications (drying between each) during daytime are recommended, boiling and scratching on the following morning.

The freshly-prepared browne of the composition given above is very slightly aqua regia-containing: a light blue liquid which after standing 3 months deposits slightly but not increasingly. Exposed to direct sunlight it gradually changes colour, with slight evolution of nitrogen tetroxide: it then slowly clears and reverts to its original blue colour, although somewhat darker, the smell of nitric oxides disappearing.

Comparative trial with a year-old "reverted" browne of this description gave an equally good final result, although more slowly: reapplication as above needed. Neither the fresh nor the reverted browne causes after-rusting.

Note 2. The roundabout method of preparation invites criticism, and is difficult to understand, as its final result of the preliminary reduces itself to plain ferric chloride, far more conveniently and cheaply bought from chemical dealer or works. "Tincture of ferric chloride" is notorious for bad keeping qualities, and the preparation of such an ocean of unstable compound, of which only 6 oz. are needed to make up nearly 2.4 quarts of browne, cannot be characterised otherwise than as most unpractical: preparation—whether for private or State Armoury use—by means of ready-made ferric chloride and the exact quantity of alcohol required, producing an exactly identical result with infinitely less trouble.

D. 11. Twist barrels. (see note to C. 10).

	a.	b.	c.	d.	e.	f.	
Mercuric chloride	3	37	11	11	18	22	grs.
Copper sulphate	23	27	11	30	18	11	"
Sol. ferric chloride 29%	23	22	33	80	69	58	"
Nitric acid D. 1.42	1	19	11	31	22	22	"
Spirit of nitre	91	110	40	40	22	137	"
Alcohol 90°	9	29	55	73	97	82	"
Water t.s. to make	¼	¼	¼	¼	¼	¼	U.S.pt.

	a.	b.	c.	d.	e.	f.	
Mercuric chloride	0.16	0.2	0.6	0.6	1.0	1.2	g
Copper sulphate	1.25	1.5	0.6	1.65	1.0	0.6	"
Sol. ferric chloride 29%	1.25	1.2	1.8	4.4	3.75	3.2	"
Nitric acid D. 1.42	0.05	1.05	0.6	1.7	1.2	1.2	"
Spirit of nitre	5.0	6.0	2.2	2.2	1.2	7.5	"
Alcohol 90°	0.5	1.6	3.0	4.0	5.3	4.5	"
Water t.s. to make	100.0	100.0	100.0	100.0	100.0	100.0	ccm

	g.	h.	k.	l.	m.	n.	
Mercuric chloride	26	27-37	46	46	64	66	grs.
Copper sulphate	26	46	23	46	40	44	"
Sol. ferric chloride 29%	22	64	62	91	73	29	"
Nitric acid D. 1.42	26	110	23	46	80	33	"
Spirit of nitre	49	274	137	82	110	66	"
Alcohol 90°	29	365	91	46	110	219	"
Water t.s. to make	¼	¼	¼	¼	¼	¼	U.S.pt.

	g.	h.	k.	l.	m.	n.	
Mercuric chloride	1.4	1.5-2.0	2.5	2.5	3.5	3.6	g
Copper sulphate	1.4	2.5	1.25	2.5	2.2	1.8	g
Sol. ferric chloride 29%	1.2	3.5	3.4	5.0	4.0	1.6	"
Nitric acid D. 1.42	1.4	6.0	1.25	2.5	4.4	1.8	"
Spirit of nitre	2.7	15.0	7.5	4.5	6.0	3.6	"
Alcohol 90°	1.6	20.0	5.0	2.5	6.0	12.0	"
Water t.s. to make	100.0	100.0	100.0	100.0	100.0	100.0	ccm

	o.	p.	q.	r.	s.	
Mercuric chloride	69	69	71	80	80	grs.
Copper sulphate	46	47	47	55	55	"
Sol. ferric chloride 29%	35	31	35	33	100	"
Nitric acid D. 1.42	35	35	35	40	35	"
Spirit of nitre	69	69	71	80	227	"
Alcohol 90°	82	111	113	44	310	"
Water t.s. to make	¼	¼	¼	¼	¼	U.S.pt.

	o.	p.	q.	r.	s.	
Mercuric chloride	3.75	3.75	3.9	4.4	4.4	g
Copper sulphate	2.5	2.6	2.6	3.0	3.0	"
Sol. ferric chloride 29%	1.9	1.7	1.9	1.8	5.5	"
Nitric acid D. 1.42	1.9	1.9	1.9	2.2	1.9	"
Spirit of nitre	3.75	3.75	3.9	4.4	12.5	"
Alcohol 90°	4.5	6.1	6.2	2.4	17.0	"
Water t.s. to make	100.0	100.0	100.0	100.0	100.0	ccm

	t.	u.
Mercuric chloride	82	82 grs.
Copper sulphate	55	110 "
Sol. ferric chloride 29%	41	46 "
Nitric acid D. 1.42	100	46 "
Spirit of nitre	82	91 "
Alcohol 90°	132	18 "
Water t.s. to make	¼	¼ U.S.pt.

	t.	u.
Mercuric chloride	4.5	4.5 g
Copper sulphate	3.0	6.0 "
Sol. ferric chloride 29%	2.25	2.5 "
Nitric acid D. 1.42	5.5	2.5 "
Spirit of nitre	4.5	5.0 "
Alcohol 90°	7.2	1.0 "
Water t.s. to make	100.0	100.0 ccm

Note. Spirit of nitre:
4% when from American sources.
2.5% when from other.

a. Greener, 1846 and 1910, for twist barrels.
b. Stelle & Harrison, do.
c, d, e. Birmingham black brownes.
f. E. C. Crossman, Springfield Armoury black.
g. "Arms & the Man".
h. "Metallarbeiter" and J. V. Howe.
k. "Metallarbeiter".
l, m. Source unrecorded.
n. C. Baker.
o. Greener, black for steel barrels.
p. C. W. Sawyer.
q. U. S. Bureau of Standards.
r. Stelle & Harrison, for twist barrels.
s. "Metallarbeiter".
t. U. S. Manual 1904.
u. Brannt & Wahl, shotgun brown.

To (a). Working instructions as for D.9.c.

To (b). Scratch and repeat until colour dark enough, scald, rinse with water slightly acidified with hydrochloric acid to bring out pattern, rinse, polish and oil.

To (c), (d), (e). Same procedure as for C.13.h. and C.14.h. Addition of mercuric chloride to (c) and (e) optional. When the browne gets "dull" it can be revived by the addition of 2.2 g per deciliter (40 grs. per ¼ U. S. pt.) of spirit of nitre.

To (f). According to Mr. Crossman, one of the brownes used in the U. S. Government Armouries. First rusting for 24 hours in steam chamber, then 2 repetitions of 3 hours each in same, boiling and scratching after each pass, finally oiling, giving an excellent deep black.

To (k). Black browne, according to Mr. Howe equal to that used by the Winchester Repeating Arms Co. Rust for 24 hours, scratch and repeat 4 to 10 times. By first starting with 2 to 3 passes with "express" browne Ac. 6 the duration of the whole procedure can be reduced to 4 days, and when thus carried out an "amazing result" is obtained.

To (n).   According to Mr. Howe the standard browne of the Savage Arms Co. The factory routine being rusting periods of 3 to 6 hours each in steam chamber, boiling and scratching after each. 3 passes are stated to give a good, deep black.

To (q).   According to C. W. Sawyer the ordinary browne for brown colour used from 1800 to 1860 by the U. S. Government Armouries, also for plain finish sporting guns. Ordinary operating method for brown colour.

To (s).   Dissolve mercuric chloride in alcohol, add remaining components, bring to volume and let stand for 4 to 6 weeks. Stated to produce a brown colour with 2, and black with 3 passes.

To (t).   Quicker in action than (k) to (m), but deposits heavily on long standing.

To (u).   Dissolve mercuric chloride in alcohol, add remaining ingredients, let stand for 24 hours in glass-stoppered bottle, and make up to volume.

Wash parts with hot water after each rusting and scratch. After 3rd pass rub with polishing leather and olive oil, adding a little turpentine, let dry for 24 hours, and finally polish with olive oil.

D. 12. Black browne.

	a.	b.	c.		a.	b.	c.
Mercuric chloride	27	38	55	grs.	1.5	2.1	3.0 g
Ammonium chloride	14	19	27	"	0.75	1.05	1.5 "
"Dissolved iron oxide"	110	137	219	"	6.0	7.5	12.0 "
Nitric acid D.1.42	27	38	55	"	1.5	2.1	3.0 "
Alcohol 90°	31	97	91	"	1.7	5.3	5.0 "
Water t.s. to make	¼	¼	¼	U. S. pt.	100.0	100.0	100.0 ccm

a. Schützelhofer.   b. Bauer & Roth.   c. Own composition.

In comparative trial browne (a) gave a very fine, deep black, but very slowly at room temperature. Concentration (c) was satisfactorily quicker, giving equally fine finish. The browne deposits after long standing: very slight tendency to after-rust.

"Dissolved iron oxide" (G. "Eisenoxyd, flüssig") is an archaic-term signifying a solution of ferric nitrate containing 5 G-% of metallic iron, that is, 21.66 G-% of anhydrous salt. Similar ancient terms in English and French.

D. 13. Black browne.

	a.	b.	c.		a.	b.	c.
Mercuric chloride	37	47	46	grs.	2.0	2.55	2.5 g
Sol. zinc chloride 50%	146	18	46	"	8.0	1.0	2.5 "
Sol. ferric chloride 29%	113	183	183	"	6.2	10.0	10.0 "
Nitric acid D. 1.42	31	47	46	"	1.7	2.55	2.5 "
Alcohol 90°	38	200	183	"	3.2	11.0	10.0 "
Water t.s. to make	¼	¼	¼	U. S. pt.	100.0	100.0	100.0 ccm

a. F. Voytier.   b. Schützelhofer.   c. Own composition.

To (a). Repeat 8 times, boiling for 5 minutes and scratching after each pass.

To (b). Zinc chloride is given by Schützelhofer as "Galmeitinktur," which according to his statement means "concentrated" alcoholic solution of zinc chloride: degree not ascertainable, but assumed as 40 G-%.

In comparative trial (a) and (c) both gave a very good, deep black on soft and hard steel, with very slight tendency to after-rusting, scarcely appreciable on hardened parts.

D. 14. Black browne (Schützelhofer).

Mercuric chloride	137 grs.	7.5	g
Zinc nitrate cryst.	150 "	8.25	"
Sol. ferric chloride 29%	91 "	5.0	"
Alcohol 90°	91 "	5.0	"
Water t.s. to make	¼ U. S. pt.	100.0 ccm	

In this formula, as taken from Schützelhofer's book, and quoted from the "Schweizerische Werkmeister-zeitung," there are unmistakable misprints or other errors, giving the wholly impossible quantity of 75 G/V-% of mercuric chloride, and an amount of nitric acid quite insufficient to dissolve the "Galmei" (calamine, zinc carbonate) in the proportion given. The above composition can therefore be only regarded at best as a guess.

Working instructions as for C. 22. After last scratching steep for 10 to 15 minutes in hot but not boiling logwood decoction.

D. 15. Black browne (Buchner).

Mercuric chloride	58 grs.	3.2	g
Zinc sulphate cryst.	58 "	3.2	"
Sol. ferric chloride 29%	58 "	3.2	"
Nitric acid. D. 1.42	58 "	3.2	"
Alcohol 90°	234 "	12.8	"
Water t.s. to make	¼ U. S. pt.	100.0 ccm	

No further details.

D. 16. Black browne (Brandeis).

Mercuric chloride	18 grs.	1.0	g
Ammonium chloride	37 "	2.0	"
Sol. ferric chloride 29%	7 "	0.4	"
Hydrochloric acid D. 1.16	37 "	2.0	"
Sulphuric acid D. 1.84	55 "	3.0	"
Alcohol 90°	77 "	4.2	"
Water t.s. to make	¼ U. S. pt.	100.0 ccm	

In comparative trial gave very good but visibly brownish black, better than that produced by Ab. 4, with very slight tendency to after-rusting, so that 1 hour's neutralisation is sufficient.

D. 17. Twist barrels (see note to C. 10).

	a.	b.		a.	b.	
Mercuric chloride	57	55	grs.	3.1	3.0	g
Silver nitrate	1	1	"	0.05	0.05	"
Sol. ferric chloride 29%	13	124	"	0.7	6.8	"
Nitric acid D. 1.42	13	16	"	0.7	0.9	"
Spirit of nitre 4%	57	55	"	3.1	3.0	"
Alcohol 90°	18	16	"	1.0	0.9	"
Water t.s. to make	¼ U. S. pt.			100.0	100.0	ccm

a. Stelle & Harrison. b. "Arms & the Man."

To (a). In the source the addition of a "small piece of chalk" is given, which would probably neutralise all the free acid. All chlorides would have to be replaced by nitrates to avoid precipitation of the silver salt, if the very small amount of the latter is indeed of practical importance.

To (b). This browne is recommended for hard or so-called "English twist". Apply and scratch 3 times daily after every 6 hours: if action is too slow, scald and start afresh. Repeat till dark enough, rinse with diluted hydrochloric acid, which brings out the figure and heightens the sheen: then scald, polish and oil.

D. 18. Twist barrels (see note to C. 10).

	a.	b.	c.		a.	b.	c.	
Mercuric chloride	18	7.5	18	grs.	1.0	0.4	1.0	g
Copper sulphate	37	15	37	"	2.0	0.8	2.0	"
Ferrous sulphate cryst.	18	7.5	18	"	1.0	0.4	1.0	"
Sol. ferric chloride 29%	23	10	26	"	1.25	0.55	1.4	"
"Black brimstone"	18	7.5	..	"	1.0	0.4	..	"
Nitric acid D. 1.42	9	4	9	"	0.5	0.2	0.5	"
Spirit of nitre 4%	55	23	46	"	3.0	1.25	2.5	"
Alcohol 90°	31	14	..	"	1.7	0.75	..	"
Water t.s. to make	¼	¼	¼	U.S.pt.	100.0	100.0	100.0	ccm

a. Stelle & Harrison.
b. Colonel Townsend Whelen.
c. Own composition.

Working instructions. Let parts rust for 24 hours, rub off loose rust with woollen rag, repeat till colour is dark enough, scald thoroughly and rub well with linseed oil. This is recommended as a very good, well-tried browne for twist barrels (these not being boiled or steamed before scratching).

Mr. Greener states (see browne D.9.c.) that the addition of "black brimstone" (crude sulphur, because really black) assists in obtaining a fine plum-brown on twist barrels. In order to ascertain whether it had a similar effect in black-colour operation, the author prepared this browne in composition (b), then in half- and full-concentration (c).

In concentration (b) and half-strength (c) its action at room temperature was very slow, in full-strength (c) satisfactorily quicker, but still very moderately so. A satisfactory deep black was obtained with both compositions in 4 to 6 passes, but the author was unable to detect any discernable difference in the trial pieces treated with or without the sulphur addition. It is not soluble in the browne in any case, and any effect it may have when in suspension is apparently merged in the general deep shade of the black finish.

Colonel Whelen states that for nickel steel about 12 passes are needed, while Poldi-Anticorro may require as many as 30.

D. 19. "Ferrooxydin" ("Metallarbeiter," Buchner).

Preparation. 10 g (154 grs.) of steel shavings are dipped in gasoline and set alight, to get rid of all grease, then dissolved, gently warming, in a mixture of 20 g (308 grs.) of hydrochloric acid (strength not stated) and 30 g (463 grs.) of nitric acid of 40° Baumé (D. 1.38 = 61 G-%).

With a portion of 100 ccm of distilled water a paste is made in a porcelain dish of 0.5 g (7.7 grs.) each of ferrous and mercuric sulphates and 2.0 g (31 grs.) of copper sulphate, and added to the solution of shavings. With the remainder of the water 20 g (308 grs.)—according to Buchner, 22.5 g—of ferric chloride solution 29% are diluted and added to the rest.

Pre-treatment (according to Buchner). Scabbards are matted by sand-blasting and scratch brush, then degreased and electrolytically iron-plated (G. "verstählt").

> *Working method.* The browne is then applied, and the parts left 30 minutes in drying oven at 80° to 100°C. (175-212°F.), 30 do. in steam oven over vapour of a water trough maintained at a temperature of 90°C. (194°F.), then boiled for 5 minutes and scratched with brush of crinkled steel wire (2 to 3/1000″) and water to which a decoction of 10 g of gall nuts per liter (145 grs. per U. S. quart) has been added. Repeating 2 to 3 times.

After-treatment. Lay parts for about 5 minutes in a mixture of 1 part linseed oil and 2 parts kerosene warmed to 100°C. (212°F.) and after cooling wipe off excess of oil with gasoline.

The coating obtained is deep ebony-black and is further heightened by adding 1 g per liter (15 grs. per U. S. quart) of powdered mountain blue (basic copper carbonate) to the scratching water instead of the gall nuts.

> *Remarks.* The above description is quoted in extenso from the source, and according to the author's experience, is an outstanding example of the way in which a browne should *not* be constituted and prepared. Theoretical considerations were amply confirmed by practical trial.
>
> The iron is in large excess and is incompletely dissolved by the mixed acids: in addition to a residue there is besides enough iron left to reduce all the ferric chloride formed by the aqua regia and added subsequently, to ferrous salt, so that further

addition of the latter is more than ever needless. The procedure of dissolving iron in aqua regia and adding ferric chloride afterwards can only be described (popularly) as "carrying coal to Newcastle," and it is hard to recognize any advantage in the substitution of the difficultly soluble mercuric sulphate for the more customary chloride.

Carrying out the prescribed mode of preparation, the result is a dark-brown, extremely muddy liquid, which in view of the exceess of iron, the easy oxidisability of ferrous salts and the absence of free acid, is continually depositing and as a browne is positively unworkable.

As this failure can only be attributed to the disproportion of certain ingredients, the author experimented with suitably modified compositions.

Starting from the prescribed quantities of acids, the weight of shavings was cut down to 5.5 g (85 grs.), the added ferric chloride and ferrous sulphate omitted, mercuric sulphate replaced by chloride, and a little free acid left over to improve keeping qualities, otherwise substantially following the original instructions.

The result was an average of 143 ccm of a dark-brown, slightly cloudy liquid, whose composition, assuming hydrochloric acid of 1.16 and 1.19 density to start with, and the admissibility of the "conventional" mean aqua-regia reaction (see under this heading, Chapter III.6, "Chemicals" after converting the ferric salts to their equivalents in standard solutions, was approximately:

Using hydrochloric acid of	(a).D.1.16	(b).D.1.19.
Mercuric chloride	0.35	0.35 g
Copper sulphate	1.4	1.4 "
Sol. ferric chloride 29%	18.75	22.15 "
Sol. ferric nitrate 28%	30.0	25.1 "
Nitric acid D. 1.42	2.03	3.6 "
Water t.s. to make	100.0	100.0 ccm

these suggested a "mean" composition containing sensibly equal amounts of (anhydrous) ferric chloride and nitrate:

(c).			
Mercuric chloride	6.4 grs.	0.35 g
Copper sulphate	25.5 "	1.4 "
Sol. ferric chloride 29%	420 "	23.0 "
Sol. ferric nitrate 28%	438 "	24.0 "
Nitric acid D. 1.42	64 "	3.5 "
Water t.s. to make	¼ U. S. pt.	100.0 ccm

The "acid-preparation" of which (on the basis of the conventional aqua-regia reaction, ante) would be:

(d).			
Mercuric chloride	6.4 grs.	0.35 g
Copper sulphate	25.5 "	1.4 "
Steel shavings	89 "	3.9 "
Hydrochloric acid D. 1.16	328 "	18.0 "
Nitric acid D. 1.42	328 "	18.0 "
Water t.s. to make	¼ U. S. pt.	100.0 ccm

For comparative trial brownes (a) and (c) were then both "salt-" and "acid-prepared" in full and half-concentration, the former being visibly very strong. The "salt-prepared" brownes are very slightly-yellowish, light green liquids, the "acid-prepared" dark brown, due to the presence of nitrosyl chloride resulting from the use of aqua regia. The "acid-prepared" full brownes deposit slightly after standing for some time, the "salt-prepared" less or not at all: in half-concentration the "acid-prepared" appear to keep better without alteration.

These brownes are now workable and indeed excellent. In comparative trial—omitting sandblasting and iron-plating—the full-brownes acted rapidly (as to be expected) on both soft and hard steel: periodical inspection of the parts being necessary to avoid pitting.

A single pass with the full-brownes gave a very acceptable and for many purposes sufficiently good black, 2 passes a very fine, deep ebony-black which did not sensibly become more intense—so far as the author's trials went—on further repetition.

The half-brownes were considerably slower in action: reapplication after drying recommended, and the final result with 3 to 4 passes equally excellent, and risk of pitting much lessened. A qualitative difference in excellence between compositions (a) and (c) was not discernable with certainty.

Both preparations show a slight tendency to after-rusting: neutralisation therefore advisable if not always unconditionally necessary.

"Acid-prepared" brownes being "keener" than "salt-prepared," the latter mode of preparation is preferable for full-brownes, their high concentration rendering them liable to cause pitting if not carefully watched. Half-brownes can be made up either way at choice, depending on the rust-resisting quality of the metal treated.

GROUP E

Brownes containing Iron and 3 other Metallic Salts

E. 1. Black browne (Krause).

Mercuric chloride	7	grs.	0.4 g
Antimony trichloride	55	"	3.0 "
Copper sulphate	37	"	2.0 "
Ferrous sulphate	91	"	5.0 "
Nitric acid. D. 1.42	20	"	1.1 "
Spirit of nitre 2.5%	18	"	1.0 "
Water t.s. to make	¼ U. S. pt.		100.0 ccm

Working instructions. Apply and let parts stand for 20 minutes in drying oven at 100°C. (212°F.). After rusting (in steam chamber) scratch with fine wire (2½ 1000″) brush and water, then boiling or steaming for 20 minutes and scratching. Repeat 2 to 3 times, dry and rub with linseed oil.

E. 2. Black browne (Michel).

Mercuric chloride	37	grs.	2.0 g
Bismuth oxychloride	37	"	2.0 "
Copper chloride	18	"	1.0 "
Sol. ferric chloride 29%	64	"	3.5 "
Hydrochloric acid D. 1.16	110	"	6.0 "
Alcohol 90°	91	"	5.0 "
Water t.s. to make	¼ U. S. pt.		100.0 ccm

This composition is qualitatively the same as Creusot Ac. 1., with the addition of ferric chloride: whether to advantage is questionable in view of the special excellence of Ac. 1.

E 3. Brown colour (Krause).

Mercuric chloride	73	grs.	4.0 g
Copper sulphate	73	"	4.0 "
Zinc sulphate cryst.	73	"	4.0 "
Sol. ferric chloride 29%	151	"	8.3 "
Nitric acid D. 1.42	55	"	3.0 "
Alcohol 90°	310	"	17.0 "
Water t.s. to make	¼ U. S. pt.		100.0 ccm

Working instructions. Apply twice in succession, let rust for 12 hours, dip for 5 to 10 minutes in boiling water and scratch with fine wire brush. Repeat until colour dark enough (generally from 2 to 3 times), then oil.

E. 4. Brown colour.

	a.	b.		a.	b.
Mercuric chloride	18	18 grs.		1.0	1.0 g
Zinc nitrate cryst.	22	22 "		1.2	1.2 "
Cream of tartar	64	64 "		3.5	3.5 "
Sol. ferric chloride 29%	64	64 "		3.5	3.5 "
Nitric acid D. 1.42	6	6 "		0.3	0.3 "
Alcohol 90°	91	33 "		5.0	1.8 "
Water t.s. to make	¼	¼ U. S. pt.		100.0	100.0 ccm

a. Dr. Winckler.
b. "Metallarbeiter." Described as "English brown." Zinc nitrate calculated from reaction between zinc carbonate and nitric acid.

Working instructions. Apply every 2 hours and scratch every 24, repeat until dark enough, scald, scratch and oil.

E. 5. Twist barrels ("Arms & the Man." See note to C. 10).

Mercuric chloride	9 grs.	0.5	g
Copper sulphate	18 "	1.0	"
Potassium nitrate (nitre)	37 "	2.0	"
Sol. ferric chloride 29%	16 "	0.9	"
"Black brimstone"	9 "	0.5	"
Nitric acid D. 1.42	2.4 "	0.13	"
Spirit of nitre 4%	37 "	2.0	"
Alcohol 90°	22 "	1.2	"
Water t.s. to make	¼ U. S. pt.	100.0	ccm

Working instructions. Apply and scratch 3 times daily at 6-hour intervals: repeat till dark enough and rinse with weak hyrochloric acid to bring out the figure. Scald with boiling water, dry and oil with linseed oil. This browne is qualitatively the same as D. 18, substituting nitre for ferrous sulphate. For effect of "black brimstone," see note to D. 9. c.

Out-E. "Express" browne containing iron and 5 other metallic salts. (J. V. Howe).

Mercuric chloride	46 grs.	2.5	g
Copper chloride cryst.	9 "	0.5	"
Sol. ferric chloride 29%	64 "	3.5	"
Potassium chlorate	61 "	3.35	"
Potassium nitrate	73 "	4.0	"
Sodium nitrate	18 "	1.0	"
Spirit of nitre 4%	137 "	7.5	"
Water t.s. to make	¼ U. S. pt.	100.0	ccm

General mode of working as for Ad. 3.

Note on "Express" brownes generally. On comparing the compositions of the "express" brownes Ab. 10, Ac. 6, Ad. 3 and "out-E," it will be noted that they are essentially combinations of mercuric chloride, chlorate and nitrate, spirit of nitre (utility questionable) and a few incidental addenda, either to meet special materials or conditions or for some unspecified (or inexistant?) purpose.

Under reservation of the effect of sodium nitrate (which the author has not tried out), they are applied in substantially the same manner and give very similar results. The complication of the above composition is striking, but Mr. Howe does not state that it gives any better result than other simpler combinations.

Under his Numbers 10, 11, 12 and 13 he lists as "express"

brownes compositions qualitatively identical with those of C. 28, Ac. 1, Aa. 22 and C. 19, respectively, each particularly good "standard" brownes, but of composition totally differing from the four named at the head of this note.

Tried out in the exact manner prescribed for "express" brownes, and scratching with a very soft iron wire brush, the author obtained completely negative results in each case: the brush swept any deposit clean, and after 6 passes the test pieces showed no vestige of any coating.

GROUP F

Combined Brownes

This group includes processes in which 2 or more brownes are used in succession or alternately.

F. 1. Silver browne (Hufschmidt).

The parts are moistened with 1% nitric acid (according to Buchner, 10%) and scratched when rusted, and this is repeated until a fine, light brown, very durable coating has been obtained.

Repeated application of a 0.2% solution of silver nitrate and exposure to direct sunlight follows, and will darken this to nearly a black colour; the layer of oxide is sprinkled through with extremely finely divided silver. Finish by rubbing with wax and hardwood polishing stick, which increases the sheen and durability of the coating.

(Dr. Schmidt, "Chemie für Metallarbeiter," gives the following proportions:

a.	Fuming nitric acid	37 grs.	2.0	g
	Water t.s. to make	¼ U. S. pt.	100.0	ccm
b.	Silver nitrate	18 grs.	1.0	g
	Water t.s. to make	¼ U. S. pt.	100.0	ccm

F. 2. Copper sulphide black (Hufschmidt and others).

The parts are dipped into a 5% solution of copper sulphate, slightly acidulated with hydrochloric or nitric acid (according to others, in a saturated solution of this salt) until they have taken on a pink colour, from deposit of metallic copper, avoiding leaving them too long in the solution, which would cause the copper deposit to become loose and powdery. Then rinse and dip into a solution of ammonium sulphide or "liver of sulphur" (potassium sulphides) until the copper deposit becomes black. Wash thoroughly with hot water, dry and finish by rubbing with polishing stick and wax.

Note. This is a very usual emergency method of colouring small bright spots on browned work (see "Retouching,"

Chapter II.8.), but is quite unsuitable for colouring entire parts, as the colour is obtained on the thin and not very strongly adherent copper deposit. Trade preparations of this nature, and others including photographers "hypo" are advertised under fancy names and at fancy prices out of proportion to their merits: they are useful for temporary touching-up but nothing more.

An improvement consists in electroplating with copper, then colouring as before, but this is still a makeshift, and at best an inferior method of colouring "stainless" barrels, Chapter II..11.

F. 3. Copper arsenite black   (Field & Bonney).

a.	Copper sulphate	37 grs.	2.0	g.
	Hydrochloric acid D. 1.16	27 "	1.5	"
	Water t.s. to make	¼ U. S. pt.	100.0	ccm
b.	Copper sulphate	11 grs.	0.6	g.
	White arsenic	410 "	22.5	"
	Hydrochloric acid D. 1.16	¼ U. S. pt.	100.0	ccm

Warm gently to dissolve the arsenic in the acid, avoiding the acid fumes. Stir parts in (a) until they have taken on a copper colour and without rinsing dip into (b), stirring all the time, until they have become smoky black. Rinse with boiling water and dry in oven or on hot plate, scratch and darken if need be by a second pass through (b), scratch and finish by lacquering with spirit, *not* cellulose lacquer.

F. 4. Sulphide black   ("Metallarbeiter").

a.	Mercuric chloride	68 grs.	3.75	g
	Copper sulphate	68 "	3.75	"
	Sol. ferric chloride 29%	32 "	1.75	"
	Fuming nitric acid	68 "	3.75	"
	Water t.s. to make	¼ U.S.pt.	100.0	ccm

b. 1.1 g "liver of sulphur" in 100 ccm of water (20 grs. in ¼ U. S. pt.).

Coat parts with (a), let dry, rub with deer leather brush and repeat until dark enough. Then lay for 20 to 30 minutes in (b), rinse with hot (soft) water, then with soapsuds, dry and rub thoroughly with linseed oil varnish.

F. 5. Cold bronze colour   ("Metallarbeiter").

Bronze bath (a).

Stannous chloride cryst.	27 grs.	1.5	g
Copper sulphate	18 "	1.0	"
Hydrochloric acid D.1.16	37 "	2.0	"
Water t.s. to make	¼ U.S. pt.	100.0	ccm

Sulphide bath (b).

Sodium hyposulphite	274 grs.	15.0	g
Hydrochloric acid D. 1.16	137 "	7.5	
Water t.s. to make	¼ U.S. pt.	100.0	ccm

Preparation (a). Dissolve the tin salt in the acid diluted with 10 ccm water, add the copper sulphate previously dissolved in 25 ccm water, and bring to volume.

After a little stirring the solution becomes cloudy and a dense flocculent precipitate forms: as soon as this has settled the bath is ready for use, even if not quite clear.

(b). Hot or boiling water alone is not sufficient to dissolve the "hypo": on account of its large latent heat of solution, continuous heating is necessary. After cooling add the acid and stir, upon which it becomes cloudy by precipitation of yellow, flocculent sulphur of molecular fineness: filtration through the closest-woven linen or finest wire gauze is needed to free it from sulphur flakes, which would otherwise cause a patchy coating.

The solution will keep usable for about 2 hours, but can to a certain extent be refreshed by further addition of hydrochloric acid and refiltration.

Working instructions. The parts are dipped for 10 seconds, not longer (but according to Hufschmidt for 15 minutes) into the bronze bath, in which they become covered with a copper-tin deposit, then rinsed and dipped into the "hypo" bath for 2 to 3 minutes, until they have become black. After rinsing and drying rub with polishing stick and wax, upon which the coating is a brilliant black.

To get a brown colour, the parts are warmed on an iron plate—open fire or flame is too uneven—to "hissing" temperature, then *very* sparingly oiled with linseed oil and again warmed until the desired shade is obtained. A certain degree of skill and practice is needed to obtain a pleasing finish, and only the thinnest film of oil must be applied.

F. 6. Black browne   (C. W. Sawyer).

For barrels difficult to rust the following procedure is recommended:

    a. First obtain a coating with browne D. 11.p.
    b. If possible, dip in, or coat with browne D. 8.a. and keep moistened with this for at least 15 minutes, then scratch.
    c. Repeat alternately with D. 11.p. and D.8.a., at least 3 times each, scratching after each pass.
    d. Finally brown with C. 14.s. (no information as to repeating).
    Note. No more precise detail is given regarding the kind of steel requiring this procedure, nor as to duration of rusting (in steam chamber or otherwise), boiling, etc.

F. 7. Black colour according to Thirault (Voytier, Schuberth).

F. Voytier states that the browning process used in the French Government Armouries is as follows: no further details.

    a. 2 passes with browne Ab.6.
    b. 5 do.   C. 10.a.
    c. 5 do.   B.5.

According to Schuberth brownes (a) and (b) are each applied twice: when rusted, the parts are laid for 5 to 10 minutes in water heated to 90° to 100°C. (194° to 212°F.), then scratched. "A few" passes with (c) follow, scratching between each, then a "long spell" (of immersion ?) of "very weak liver of sulphur solution" (not mentioned by Voytier). After drying, again several passes with (c) in continually increasing dilution, then rubbing with olive oil, wiping, dipping in water at 60° (140°F.), drying and rubbing with oil.

> No further details than this rather fragmentary instruction are given in the source, which is not distinguished for precision.

GROUP G

Collective Browning

This group is characterised by the immersion of the parts to be coloured in a bath, colouring being effected by simple boiling of the work-pieces. A number of parts can thus be treated together instead of each singly in turn.

As the colouring takes place evenly over the entire surface, hollows which the scratch brush cannot touch are browned along with the rest: where this is not desired, the surfaces to be left bright must be protected by a suitable varnish, its composition depending on the nature of the colouring bath.

Collective browning is specially suited to smaller gun parts, tools, automobile and cycle accessories, for which regular browning would be too troublesome or expensive.

G. 1. Cold nitric black   (Vannino & Mauermayer).

Smaller parts are coloured deep black in a few minutes by immersion in concentrated fuming nitric acid, which must be kept as cold as possible to avoid loss of lower nitrogen oxides. Complete degreasing essential and longer period of dipping unnecessary.

Then rinse thoroughly, lay for a few minutes in a weak solution of caustic potash or soda, wash off all alkali, dry with rags or in hot sawdust, oil and rub with woollen cloth, which will bring up the deep black colour.

The coating adheres strongly, does not flake off, and resists concentrated nitric acid, concentrated and diluted caustic alkalies, but not diluted acids. The process is based on the so-called "passivity" of iron in concentrated nitric acid.

G. 2. Dull black   (Hartmann, Böttger).

The parts are dipped for some minutes in 8 to 10% nitric acid, stirring continuously, rinsed with boiling water, then boiled for 3 successive periods of 5 minutes each, in fresh water each time, to remove the last traces of acid. Then warm on wire gauze until the pieces are dark blue, and quench in oil: the surface will then be dull black on account of the slight pickling. Thorough riddance of the last traces of acid is essential, otherwise unsightly reddish spots and streaks remain.

G. 3. Ferricyanide blue   (Hartmann, Böttger).

Sol. ferric chloride 29%	17 to 31 grs.	0.9 to	1.7 g	
Potassium ferricyanide	6 " 9 "	0.3 "	0.5 "	
Water t.s. to make	¼ U.S. pt.	100.0	100.0 ccm	

The pieces are dipped for a few seconds into the solution (longer dipping causes partial decomposition with greenish-blue deposit) dry on hot iron plate and repeat 4 to 6 times. A fine, deep blue to black is formed, but which is partly easily rubbed off. Finish by oiling.

> *Note.* Potassium ferricyanide, *red* prussiate of potash (L. kalium ferricyanatum, F. "prussiate rouge de potasse," G. "rotes Blut-laugensalz") is distinguishable from *yellow* prussiate of potash (potassium ferrocyanide) used for casehardening, by its deep red colour and smaller crystals. Its aqueous solution gradually undergoes self-decomposition (promoted by direct sunlight), so that it must be prepared fresh just before using. Red prussiate is more poisonous than the yellow salt.

G. 4. Hyposulphite black   (Hiorns).

The parts are dipped for 10 minutes in a hot solution of 25 g of "hypo" in 100 ccm of water (457 grs. in ¼ U. S. pt.) — heating continuously when making up—dried and scratched. Repeat until obtaining a blue-black coating with warm, copper-coloured shade, for which 2 passes are usually sufficient. A stronger solution is of no advantage.

> *Note.* Sodium hyposulphite or thiosulphate (L. natrium hypo-sulfurosum, G. "unterschwefligsaures Natron") is the "hypo" used in photography for fixing. A large amount of heat is absorbed (becomes "latent") when it is dissolved, a property here troublesome, but usefully applied to the preparation of freezing mixtures.

G. 5. Lead-glance blue

	a.	b.	c.	a.	b.	c.
Lead acetate	32	42	69 grs.	1.75	2.3	3.8 g
Sodium hyposulphite	128	42	227 "	7.0	2.3	12.4 "
Water t.s. to make	¼	¼	¼ U.S. pt.	100.0	100.0	100.0 ccm

a.   "Metallarbeiter."   b.   Puscher, Hiorns.   c.   Buchner.

Dissolve the salts separately, the "hypo" as before, heating continuously, then mix.

Dip the work-pieces in the solution, maintained at a temperature of 60° to 80°C. (140° to 175°F.), stirring continuously to ensure uniform action: the colour develops rapidly, and according to Hiorns, as follows:

1. Light blue.
2. Dark brown.
3. Blue mixed with purple.
4. Blue mixed with purple but lighter colour.
5. Fine uniform light blue.
6. Steel grey.
7. Black, after about ½ hour.

When stopped at exactly the right moment, the pieces can be coloured exactly like temper colouring produced by simple heating in the open or in a sulphur bath. The bath must not be heated to boiling, as the colours then form too quickly and are more difficult to control.

> *Note.* Lead acetate (L. plumbum aceticum), popularly "sugar of lead" (same popular designation in French and German) is a poisonous salt which has acquired this name from its sweetish taste.

G. 6. Picric brown   (B. Guerini, D.R.P. 292,603/15).

16 g of picric acid and 84 g of caustic soda are dissolved in a porcelain basin in 100 ccm of water (292 and 1530 grs. in ¼ U.S. pt.) and gradually heated to 150°C. (300°F.).

The polished and degreased parts are left in the bath for about 10 minutes, until they have taken on a fine, brilliant brown, then rinsed, dipped in a mixture of oil and kerosine and dried in sawdust.

> *Note.* Picric acid (L. acidum picricum), trinitrophenol, bright yellow needle-like crystals of extremely bitter taste and poisonous, is the well-known high explosive known in different countries as melinite, "Granatfüllung 88," lyddite, Ekrasit, shimose, etc.

G. 7. Picric black   (W. Utendörfer, D.R.P. 368,548/22).

The addition of gallic acid to the sodium salt bath of nitrophenols produces an intensely black instead of brown colour on iron and steel pieces.

The picric acid bath is however easily spoiled by unskilled handling, and in consequence of its tendency to self-explosion, which is only slightly mitigated by the gallic acid addition, successful operation is conditional on adherence to strict working instructions. This is a dangerous solution to handle.

Under supplementary patent 391,800 the dangerous tendency of nitrophenols is overcome, and the bath rendered insensitive, by the addition of 2 to 5% of saccharides or polysaccharides, e.g. treacle or similar sugar-containing substances, to the nitrophenols before making up the bath, with the further advantage of increasing both the durability and the brilliance of the coating.

G. 8. Lead oxide black (Rondelli, Brit. P. 137,436/20. D. R.P. 347,934/20).

25 g of lead oxide, litharge (L. plumbum oxydatum, Fr. "massicot," "cendre de plomb," G. "Bleiglätte"), are dissolved in 1 liter (365 grs. in 1 U.S. quart) of caustic soda lye D. 1.38 (this contains 35 G-% or 48.3 G/V-% of alkali and boils at 120°C.=250°F.).

The degreased pieces are boiled in this, in which they become covered after a few minutes with a brilliant, non-adhering coating of metallic lead crystals. In about $\frac{1}{2}$ hour the colour has attained its maximum density.

The lead powder is wiped off, the parts thoroughly washed with boiling water, dried and dipped for a few minutes in boiling, rather thin mineral oil (D. abt. 0.922, b.p. 345-350°C.), then wiped, leaving a fine deep black. The excess of lead remains in the bath and can be recovered. The process is based on electrolytic effect, but without requiring any current from an external source.

As the colour is produced in an alkaline medium, after-treatment is limited to simple rinsing: there is no after-rusting and the coating does not peel.

> *Note.* This so-called "Brondsé" process is extensively used in black-colouring smaller articles of mass-production, such as cycle, typewriter or automobile accessories, spanners, etc. and small parts of complicated shape such as rear and peep sights, to which it is excellently suited. It is not suitable for barrels and large parts like receivers, the rusting-cum-scratching procedure which secures uniform colour here being absent.

G. 9. Cupro-silver black (Schuberth).

The pieces are dipped for 10 minutes into a bath of 20 g each in 100 ccm of water (365 grs. in ¼ U.S. pt.) of copper and silver nitrates, rinsed and warmed in sand bath until the desired, finally deep black colour appears. Expensive but excellent for small parts.

G. 10. Arsenic grey (Hiorns).

Boil 11 g of white arsenic (L. acidum arsenicosum, arsenic trioxide, Fr. popularly "mort aux rats," G. ditto "Giftmehl") in water with enough strong hydrochloric acid to effect solution, and bring volume to 100 ccm (200 grs. arsenic in ¼ U.S. pt.). The solution is without action cold, but parts boiled in it take on a fine grey colour.

G. 11. Arsenic grey (Dr. Langbein).

Copper chloride cryst.	73 grs.	4.0 g
Sol. ferric chloride 29%	985 "	54.0 "
White arsenic	18 "	1.0 "
Hydrochloric acid D. 1.16	410 "	22.5 "
Water t.s. to make	¼ U.S. pt.	100.0 ccm

Let solids stand in the acid for 24 hours or completely dissolved, then bring to volume. The pieces are boiled for 15 minutes in the solution, then dried in sawdust.

G. 12. Arsenic grey-black (Dr. Beutel).

Ferrous sulphate *or* ferric chloride cryst.	164 grs.	9.0 g
White arsenic	146 "	8.0 "
Crude concentrated hydrochloric acid	¼ U.S. pt.	100.0 ccm

Let the mixture stand for a few days until all is dissolved, keeping in a stoppered bottle and as cool as possible.

The degreased parts when dipped or painted with the solution become covered after about ½ minute with a uniform, delicate dove-grey coating: rinse thoroughly and repeat until the desired shade is obtained. Wash in hot soda solution to destroy all traces of acid, then with plain water. The surfaces are but slightly pickled: on matted pieces the coating is velvety smooth, on polished ones a brilliant deep black. Finish by lacquering with spirit (*not* cellulose) lacquer.

G. 13. Browne for RNO steel (Communicated by MM. Auguste Francotte).

Copper sulphate	7 grs.	0.4 g
Sol. ferric chloride 29%	640 "	35.0 "
Hydrochloric acid D. 1.16	9 "	0.5 "
Nitric acid D. 1.42	33 "	1.8 "
Water t.s. to make	¼ U.S. pt.	100.0 ccm

Dip parts for 2 minutes in browne, rinse, lay for 15 minutes in 6% sulphuric acid, rinse thoroughly, dry, rub with woollen cloth and scratch with file card: repeat until the coating of oxide is a brilliant deep black.

G. 14. Sodium peroxide black (Dr. Th. Mauermeyer).

As much sodium peroxide as will lie on the point of knife is stirred into 50 ccm of very concentrated caustic soda lye (abt. 50%) and the parts boiled in it for a few minutes (b.p. 140-150°C.=280-300°F.). The metal becomes black, although not very deep, while oxygen is freely given off. Then rinse, dry and rub off with cloth or brush.

The coating does not peel, the polish of the parts is not affected, and the coating affords a certain degree of rust-protection.

G. 15. Caustic soda browne (Dr. A. Mai, D.R.P. 375,-198/18).

Alkaline baths with added oxidants for browning soon lose their effectiveness, as the carbonate formed by absorption of atmospheric carbonic acid hinders the oxidation of the metal. The products of reduction and decomposition accumulate, so that the oxidising power of the bath is soon exhausted.

By adding slaked lime to the bath the caustic alkali is continuously regenerated, promoting oxidation, and the formation of carbonate, with its harmful accompaniments, restrained.

Most inorganic oxidants can be used, such as dichromates, permanganates, nitrates, peroxides and chlorates, also aromatic nitro compounds. According to concentration and nature of metal deep black to bronze- and brass-coloured coatings can be obtained.

Mode of working. Polished iron parts can for instance be coloured black in 20 to 30 minutes, according to their size, in a bath of 100 parts of 33% caustic soda lye, heated to 120-130°C. (250-270°F.) to which 3 parts of nitre and a little slaked lime have been added, and this process can be carried on continuously. Pre-pickling the parts in a mineral acid, e.g. hydrochloric acid, deepens the colour.

By increasing the concentration of the lye and the amount of nitre, and heating to 180-200°C. (350-390°F.) polished iron parts can be coloured in about 5 minutes.

After removing from bath, rinse with water, then with a lime solvent, e.g. diluted acetic acid, to get rid of adhering lime, then several times with hot water, dry and finish with linseed oil or lacquer.

The advantage of the process is complete preservation of the surface and polish in the single operation, and regeneration of the bath by the lime addition. The oxide coating does not peel off.

Supplementary patent D.R.P. 376,669/18. It is found that a freshly prepared bath does not start and act so easily and thoroughly as one containing reduction products of the oxidants used, such as nitre, potassium permanganate, potassium or sodium dichromate, etc., that is, nitrite, manganite, chromate, etc.

If small quantities of such reduction products are added from the start, the colouring action sets in at once and produces deeper shades, an effect which can be further reinforced by the addition of small quantities of an iron salt, e.g. ferric nitrate.

If for instance 100 g of sodium nitrite and 50 g of ferric nitrate (1540 and 770 grs.) are added to a bath containing 100 kg (220 lbs.) of soda lye and 2 kg (4.4 lbs.) of nitre, darker, brass-yellow to black colours are quickly obtained. Over-energetic action on the parts can be avoided by a protective coating, e.g. by dipping into very diluted copper sulphate solution or temper-blueing, thereby making the action more gradual. It is advantageous to work with highly concentrated lye (up to complete saturation) and at temperatures of 180° to 200°C. (360-390°F.).

G. 16. Chrome browne (Metals Protection Corporation, Br. P. 264,788/26.).

Chromic acid	420 grs.	23.0 g
Sulphuric acid D. 1.84	27-200 "	1.5-11.0 "
Water t.s. to make	¼ U. S. pt.	100.0 ccm

The parts are dipped for about 5 minutes in the browne, warmed to 40°C. (105°F.), dried, rinsed, dried in oven and oiled. A yellowish to chocolate-brown, strongly adherent coating is formed, very resistant to acid fumes, e. g. of hydrochloric and nitric acids, damp-saturated air and atmospheric agents. If the parts are warmed to 100°C. (212°F.) for 15 to 30 minutes the coating takes on a greenish tint and becomes still more strongly adherent.

> *Note.* Chromic acid, more correctly chromic anhydride, CrO_3, (L. acidum chromicum) is a strongly oxidising body, crystallising in bright red, deliquescent needles. The "technical" acid contains about 60% of pure anhydride, and invariably some free sulphuric acid, the "pur. cryst." quality about 80 to 85% of chemical pure product.

G. 17. Permanganate browne for "Stainless" steel (Thos. Firth & Sons Ltd.).

The process consists in slight pickling, followed by treatment in a permanganate bath.

Pickle.   Basic bath A.

Ferric chloride cryst.	610 grs.	33.3 g
Hydrochloric acid D. 1.16	0.5 ccm	0.5 ccm
Nitric acid D. 1.42	1.0 "	1.0 "
Water t.s. to make	¼ U.S. pt.	100.0 "

Bath B.   Equal volumes of (A) and water.

Bath (C).   1 volume of (A), 4 volumes of water.

Permanganate bath.

Potassium permanganate	145 grs.	10.0 g
Sol. silver nitrate 5%	1 ccm	1.0 ccm
Acetic acid 30%	1 "	1.0 "
Water t.s. to make	1 U. S. quart.	1 liter.

The permanganate bath is used boiling, and evaporated water must be made up as necessary.   It attacks both wood and iron, so use a trough of flawless enamelled iron or stoneware.

Process.   The barrels are laid for a few seconds in the B-pickle, then rapidly and evenly wiped over with a rag dipped in the same pickle, rinsed under running water and wiped. While still wet they are dipped in the C-pickle for 15 to 30 seconds, adherent liquid smartly shaken off, and at once immersed in the permanganate bath.

After 20 to 45 seconds they are taken out and freed from adhering brown deposit with a rag, under running water: the metal has now taken on a dull look.

Scratching is now done with a soft wire brush, the barrels again dipped for 30 to 45 seconds, wiped and scratched, repeating as often as necessary.

At each pass the coating becomes darker and soon assumes a dark reddish-brown shade, and by longer boiling and after repeated passes colours up to deep black can be obtained. When dark enough the barrel is not scratched again, but washed clean, dried and oiled.

> *Note.*   In the permanganate bath decomposition and action on the metal produce brown spongy deposits of manganese oxide on the barrel, contact with which causes patchy brownish-yellow to black streaks.   For this reason it is advisable to have 2 permanganate baths, reserving one for first boiling, the second for the longer passes lasting over a minute.   The brown precipitates settle after about an hour, and can be got rid of by decanting or syphoning off the clear liquid.

The bath gradually loses its effectiveness with use, and can be revived by adding a few drops of the silver nitrate solution and acetic acid.

The barrels should not be handled by means of tongs, hooks, etc., of different metal on account of electrolytic action.

Complete sealing of the bore is specially important, otherwise the corrosive fluid penetrates and does damage. The muzzle plug must be tightly driven in and is best renewed at each pass: the chamber is best closed by a plug carrying a glass tube bent upwards at right angles to beyond the water level, so as to permit free expansion of the contained air, which would otherwise tend to loosen the plugs.

Remarks. The author had a few barrels made of Firth-Brearley stainless steel by an eminent gun manufacturer and browned by this process: some of these came out with a good deep black, the surface slightly matted: others were more or less imperfectly striped or variegated, with occasionally recognisable pitting. In one case the permanganate had penetrated the bore, leaving a black deposit which could not be got rid of by any mechanical or chemical means.

The relatively low durability of this and otherwise produced browns on stainless steel has already been noticed in Chapter II. under this heading.

G. 18. Chrome browne for Stainless steel

A later process worked out by the same firm is based on the action of a sulphuric chromic acid bath at gentle heat, of which two compositions are given.

	a.	b.		a.	b.
Chromic acid anh.	607	304 grs.		33.3	16.7 g
Sulphuric acid D. 1.84	425	914 "		23.3	50.0 "
Water	792	607 "		43.4	33.3 "

Bath (a) is for "stainless" (about 12 to 15% of chromium), (b) is for "staybrite" steel (about 18% chromium and 8 to 40% nickel).

For "stainless" the bath should be warmed to 30° or 40°C. (85-105°F.), producing a light bronze-coloured coating after 72 hours. "Staybrite" is best treated at room temperature, but action is more rapid at 30° to 50°C. (85-120°F.), but the latter temperature must not be exceeded.

G. 19. Chromic acid browne (W. H. Hatfield & H. Green, Br. P. 275,781/26).

The colouring of "rustless" steel containing up to 15% of nickel and 8 to 40% of chromium is effected in a bath containing 200 g of solid chromic acid, 400 to 450 g of sulphuric acid D. 1.84 and 1000 g of water (7.14, 16 and 35 oz. avoirdupois), warmed to 40° to 60°C. (105-140°F.), which in the course of a few hours produces an even, very satisfactory dark coating. Polished parts take on a very dark shade recalling black enamel, slightly etched ones a dark, dull colour. In general no further treatment it required than rinsing and drying. At least 150 g each of chromic and sulphuric acids must be present per liter of bath (1500 grs. per U. S. gallon).

G. 20. Selenious brownes (Malherbe).

The action of these is based on that of copper selenite, which although itself insoluble in water, dissolves on addition of nitric or sulphuric acid. The brownes are best prepared with copper sulphate and selenious acid, adding nitric acid to prevent precipitation. The following compositions are given:

	a.	b.	c.	d.
Copper sulphate	183	183	37	18 grs.
Selenious acid	110	183	11	11 "
Nitric acid D. 1.42	73-110	73-110	12 drops	9 "
Water t.s. to make	¼	¼	¼	¼ U. S. pt.

	a.	b.	c.	d.
Copper sulphate	10.0	10.0	2.0	1.0 g
Selenious acid	6.0	10.0	0.6	0.6 "
Nitric acid D. 142	4.0-6.0	4.0-6.0	10 drops	0.5 "
Water t.s. to make	100.0	100.0	100.0	100.0 ccm

On dipping the parts into the solutions they soon become covered with a loosely adhering dark-coloured coating containing copper and selenium. Washed first with water and then with alcohol and quickly dried over a gas flame, the deposit adheres strongly and after rubbing with a woolen cloth takes on a bluish to brilliant black shade.

To (a). The browne containing 4.0 g. nitric acid first deposited a brown, then a white precipitate, but which redissolved on adding the further 2 g of acid. After 2 to 3 passes as described the author obtained a beautiful deep blue coating. When used as an ordinary browne comparative trial resulted in a good, deep black.

To (b). On prolonged immersion this bath is stated to produce, in order, a yellow, pink, purple, violet and brownish-black coating, but it deposits heavily, the parts becoming quickly covered with a flocculent precipitate.

To (c). This also deposits heavily: immersion in the warmed solution is stated to produce a very fine, dark greyish-black colour.

To (d). This composition is also given by E. S. Whittier as a regular browne: rusting in steam chamber. Stated to be slow-acting, but to give very fine coatings.

Note. On account of the comparatively rare use in current practice of selenious acid (L. acidum seleniosum, H_2SeO_3), the author has only tried out composition (a), then getting exactly as good results on omitting as on including rinsing with alcohol.

GROUP H

Miscellaneous Colouring Methods

H. 1. Temper-blue. ("Metallarbeiter" & many others).

This colouring method is of ancient date and consists in superficial oxidation of the surface by heating in air, a lighter or darker colour being producible at will by regulation of temperature. The following average table can be taken as a guide:

Temperature.		*Colour.*
°C.	°F.	
220	428	Pale straw.
228	443	Straw.
232	450	Dark straw.
254	489	Light brown.
265	509	Purple red.
280	536	Violet.
285	545	Light blue.
293	560	Dark blue.
316	601	Blue-black.

It must however be borne in mind that the relation between temperature and colour is by no means rigid, the duration of heating having considerable influence. Within certain limits, a particular colour can be obtained equally well by a long exposure to a lower temperature as by a short exposure to a higher. The physical state of the surface and the chemical composition of the metal weigh largely in the scale.

The following methods of temper-colouring are in ascending order of practical excellence:

1. On iron plate over coke or charcoal fire (small parts on wire gauze over Bunsen burner).
2. In sand bath.
3. In a metallic bath just kept fluid, and of composition such that its melting point corresponds to the temperature desired. Average compositions are as follows:

Lead %	Tin %	Melting point.	
64	36	220°C.	428°F.
67	33	228	443
68	32	232	450
78	22	254	489
83	17	265	509
92	8	280	536
94	6	285	545
96	4	293	580
97	3	316	601

The parts must be kept in the bath until they have taken the desired colour, turning them about continually, then quickly cooled, dried and oiled. Colours formed slowly are as a general rule more durable, if only slightly, then those produced quickly, and their gradual appearance is more easily controlled.

Temper colouring small parts, such as screws, pins, small springs, etc., offers no difficulty, but larger parts, such as pistol and revolver frames and even rifle barrels, demand much skill and patience: it is by no means rare to find the products even of first-class manufacturers, otherwise irreproachable, quite unpleasingly blued.

Temper blueing has the advantage of being much quicker and requiring much less handwork than browning, which sufficiently explains the general preference for this finish in the case of small parts of plain or complicated form, and in the U. S., on account of high labour costs, for pistol and revolver parts generally.

The exceptionally beautiful, deep blue-black finish characterising the products of the world-renowned firm of Smith & Wesson, is produced by a special process, for the particulars of which the author has to thank the kind courtesy of Major D. B. Wesson, Vice-President of this Company. This is as follows:

A shaft revolving slowly in a gas heated cylinder carries several reels designed to receive the various parts: these reels consist of two cheeks joined by cross-bars carrying studs and pins. Frames are screwed onto threaded studs by the barrel hole, cylinders slipped onto suitable pins on the cross-bars. The drum turns in the opposite direction to the reels and contains a quantity of burnt bone and a certain precentage of a pine tar preparation called "Carbonia" (sold by the American Gas Furnace Co., Elizabeth, N. J.): as drum and shaft rotate the mixture is continually lifted and drops on the parts.

The colour is the result of the careful combination of mixture, temperature and time of treatment, determined by experience, the temperature being maintained between 320° and 370°C. (610 to 700°F.) according to the blue or nearly black shade it is desired to obtain. When the correct time has elapsed the heat is cut off, and after standing a little the still warm parts are cooled off in oil and degreased with gasoline or other suitable solvent, which completes the process.

Apart from the carefully evolved procedure detailed above, the beauty of the mirror-smooth, enamel-like finish characterised by its popular name, "Smith & Wesson blue," depends chiefly on the perfection of the polish given to the parts. The process evolved by the firm consists of 3 successive stages, rough, fine and lustre polishing: if the last be omitted the result is the ordinary grey-blue of market-ware.

During the Great War the process was modified by substituting matt-brushing ("frosting") for the final polishing stage. This consists in brushing the parts with circular brushes of soft steel wire (4/1000″) driven at emery-wheel speed (85 ft./sec. = 26 m/sec.), so that the points of the wires impinge, pine-needle fashion, on the surface of the pieces (see under "Matt-brushing," Chapter V.). The result is a very delicate, refined-looking, non-reflecting, velvety finish—generally known by this name—unquestionably superior to lustre finish from the shooting-technical point of view. All the same, the buying public clamoured for the brilliant "Smith & Wesson blue" after the War, to which the makers have reverted accordingly, on the sound commercial principle of supply and demand.

> *Note.* As proved by the above example, temper-blueing effected by a well thought-out, perfectly developed process—scarcely possible to the individual—can produce a finish which for beauty can rival the finest luxury browning.
>
> In general however, temper-blueing can only be recommended for secondary parts, as the film of oxide, although of closer grain, is of extreme tenuity, is thus more sensitive to attrition, mechanical action and weather, and in the matter of rust protection even less effective than regular browning.

H. 2. Nitre blue.   ("Shooting & Fishing," abt. 1898).

At Springfield Armoury a mixture of saltpetre (potassium nitrate) and 8 to 10% manganese peroxide is thoroughly mixed and heated to about 200°C. (390°F.), at which temperature a pinch of sawdust dropped onto it briskly burns off. The degreased parts are stirred about in it for 5 minutes, or until the desired colour is obtained, freed from adherent saltpetre with hot water, darkened in hot whale oil, cleaned with gasoline and finally oiled.

According to Hartmann, a temperature of 315°C. (600°F.), that is, about the melting point of saltpetre, is necessary for producing a deep blue colour: the same author (and many others) state that the addition of manganese peroxide is of no special advantage, save a slight quickening of the process.

H. 3. Nitrite black.

According to various authors a temper-black can be obtained in a few minutes by immersing the parts in a flux composed of 80 parts caustic soda and 20 parts sodium or potassium nitrite, then washing and oiling. The coating is stated to be durable, but liable to after-rust.

> *Note.* Avoid confusing sodium nit*rite* (L. natrium nitrosum) with sodium nit*rate* (L. natrium nitricum), that is, Chili saltpetre.

H. 4. "Orthoman" colouring.   (Buchner).

Dr. Aug. Prettner, D.R.P. 298, 207/15, states that iron and steel can be coloured by stirring in a flux consisting of alkaline chromate and chromic acid, or of alkaline polychromates, producing a coating of mixed magnetic oxide and chromic oxide. Such fluxes easily evolve oxygen, on which their action is based, so that over-heating is to be avoided: by careful temperature regulation colours from light blue to deep black are obtainable. Preparations known by the trade name of "Orthoman I & II" are (or were) available, of which the first begins to act at a temperature of 330°C. (abt. 630°F.), the second from 370°C. (abt. 660°F.) onwards; within certain limits a given shade can be obtained by a short immersion at higher temperature or vice versa, without exceeding the limit of 530°C. (940°F.). Dullfinished parts take the colour in a few minutes, highly polished ones require a little longer time. No after-treatment is required beyond washing, and for Orthoman I a little brushing.

> *Note.* Inquiries for Orthoman addressed to the source indicated by Buchner elicited no reply. All colouring methods by fused salts are obviously only applicable conditionally to parts that the requisite temperature will not injure, but are quite unsuitable to soldered or brazed assemblies, therefore of restricted use.

H. 5. Chrome patina.   ("Metallarbeiter").

The parts warmed to 100°C. (212°F.) are dipped into a 10% solution of potassium dichromate (about saturated at average room temperature), which then dries in a few seconds. They are then heated for a few minutes—preferably over a charcoal fire—and these operations repeated alternately until the colour is dark enough. Adherent salt is got rid of by washing with hot water, and the parts oiled to finish.

If stopped at exactly the right temperature, a very dark brown, iridescent coating with metallic lustre results: if insufficiently heated, the wash water is coloured yellow, and at higher temperature a dull-black, non-iridescent coating is obtained.

> *Note.* This easily and quickly obtained patina is particularly recommendable for small parts, especially those for which, like gun sights, a dull finish is altogether preferable. This method has been used by the author for many years with complete satisfaction.

H. 6. Iron patina. ("Metallarbeiter").

If instead of the dichromate solution one composed of

Ferrous sulphate cryst.	457 grs.	25.0 g
Ammonium chloride	46 "	2.5 "
Water t.s. to make	¼ U. S. pt.	100.0 ccm

is used in exactly the same way as H. 5, the treated parts acquire a more or less brilliant coating of magnetic iron oxide, according to the physical state of the untreated surfaces.

H. 7. Yellow patina. ("Metallarbeiter").

A brass-yellow coating is obtained in an exactly similar manner by the use of a solution of

Copper sulphate	9 grs.	0.5 g
Stannous chloride (tin salt)	9 "	0.5 "
Hydrochloric acid	Just enough to clear.	
Water t.s. to make	¼ U. S. pt.	100.0 ccm

H. 8. Copper oxide black ("Metallarbeiter" and others).

The parts are dipped into a nearly saturated solution of copper nitrate, heated until decomposition of the nitrate, repeating this treatment until a sufficiently dark coating has been obtained, consisting of magnetic oxide mixed with copper oxide.

To this end 100 to 125 g (1825 to 2280 grs.) of copper nitrate are dissolved in about 40 ccm (50 ccm) of warm water, and the solution brought to 100 ccm (¼ U. S. pt.). Dr. Beutel gives in addition 0.4 g (7.5 grs. in ¼ U. S. pt.) silver nitrate. In the stronger concentration the solution partly solidifies on cooling to large needles and tablets, but easily redissolves on standing in warm water: if only 100 g of nitrate are used it remains liquid.

The parts are painted with this solution and warmed on iron plate or wire gauze, continually turning them over, until the nitrate decomposes (at abt. 250°C. = 480°F.), giving off noxious nitrous fumes: then let pieces partly cool, dip and heat alternately (3 to 6 times) and brush with bristle brush: according to the number of passes a lighter or deeper black coating is obtained. If the scratching is more energetic, e. g. done with brass or very soft iron wire brush, the coating gradually takes on a brass-yellow shade similar to that produced by the manganese nitrate treatment. No after-treatment required.

(According to Dr. Buchner 100 g of copper nitrate are dissolved in alcohol and the solution brought to 100 ccm with alcohol: with this the author obtained a coating of the same colour, apparently somewhat more resistant to the brush. This solution however begins to deposit after standing for some time, so that in the end its advantage is small.)

H. 9. Manganese bronze-brown.   ("Metallarbeiter").

Exactly the same treatment with a solution of 125 g manganese nitrate in 40 ccm water, then brought to 100 ccm, produces a fine brass-yellow, for which from 2 to 3 passes are generally sufficient: if heated for a longer time a bronze-like shade results. (Other composition nitrate 70, alcohol 30 g, water to make to 100 ccm = 1280 and 547 grs. to U. S. ¼ pt.)

Intermediate colours can be obtained by mixtures of the copper and manganese solutions: equal volumes for example produce after brushing a colour almost as dark as the copper solution alone, but after scratching with a soft iron wire brush a dark bronze shade almost identical with that given by the manganese solution after prolonged heating. By varying the proportions of the mixture, the duration of heating, the number of passes and the after-treatment, all shades from pale brass to black can readily be obtained. No oiling is required.

> *Note.* The author has often had recourse to these two methods, with very satisfactory and pleasing results.

H. 10. Grey arsenical pickle.

According to Dr. E. Beutel a fine dove-grey coating is obtained by a pickle consisting of

White arsenic	146 grs.	8.0 g
Ferric chloride cryst. *or*	164 "	9.0 "
Ferrous sulphate cryst.	same	same
Crude hydrochloric acid	¼ U.S. pt.	100.0 ccm

Keep the pickle in a glass-stoppered bottle in as cool a place as possible: the arsenic dissolves slowly and the bottle should be frequently shaken, and the solution is complete only after some days.

The degreased parts are dipped for ½ to 1 minute in the pickle, rinsed, dried, and the operation repeated until a deep enough shade has developed. Then rinse, first with water, then with soda solution (to neutralise all acid residue) and finally with water: then dry and lacquer with spirit, *not* cellulose lacquer, which gives unsightly results on arsenically treated parts.

The following arsenical pickles are recommended by the Physikalisch-Technische Reichsanstalt, Berlin, as well-tried:

Light grey colour:		
Ferrous sulphate cryst.	152 grs.	8.3 g
White arsenic	152 "	8.3 "
Hydrochloric acid (1).	¼ U.S. pt.	100.0 ccm

Grey-black:

Antimony trichloride	55 grs.	3.0 g
White arsenic	110 "	6.0 "
Forge scale	274 "	15.0 "
Hydrochloric acid (1).	¼ U.S. pt.	100.0 ccm

Note. "White arsenic," more correctly arsenious anhydride (L. acidum arsenicosum) is the well-known violent mineral poison.

"Crude" hydrochloric acid has a density round about 1.13, averaging about 24 G-% of hydrochloric acid gas—and many impurities.

"Forge scale" is necessarily of variable composition, but usually consists of about ¾ ferrous and ¼ ferric oxide, corresponding approximately to the formula $Fe_2O_3 6FeO$.

(1). Acid given in source as "concentrated," but degree (or density) unstated.

H. 11. Arsenical bright pickle. (Buchner).

Antimony trichloride	73 grs.	4.0 g
White arsenic	18 "	1.0 "
"Bloodstone"	146 "	8.0 "
Alcohol 90° t.s. to make	¼ U.S. pt.	100.0 ccm

Warm solution on water bath for about ½ hour, by which time the solids will have partially dissolved. Rub the parts with a cotton rag dipped in the pickle until the colour appears, which will be a bright silvery one on highly polished pieces. Then lacquer with spirit varnish.

Note. By "bloodstone" (Fr. "pierre sanguine," G. "Blutstein") washed ferric oxide is here meant, not the natural mineral (hematite, ferric oxide).

H. 12. Fruit-acid black. (Buchner).

The discolouring of knife blades after cutting apples and similar fruit caused this chemist to compose the following browne, containing only organic acids:

Tannic acid	4 grs.	0.2 g
Tartaric acid	4 "	0.2 "
Water t.s. to make	¼ U.S. pt.	100.0 ccm

Its application is extremely simple: paint on parts, let dry and repeat until a fine, deep blue-black colour has been obtained.

Note. The author arrived at the result stated, for which many passes—25 or more—were needed. All warming, even that caused by standing in the sun, is to be avoided, otherwise the coating adheres slightly and is easily wiped off, but if formed throughout in the cold it is relatively strongly adherent. After getting a sufficiently dark shade the parts are simply wiped and oiled, without any brushing.

The method is of elementary simplicity and gives very acceptable results, but the durability of the coating, especially against attrition, is relatively low, and at best equal to that resulting from the "oil-blacking" process used in emergency for gun parts.

H. 13. Acid-proof bronze-brown.   ("Metallarbeiter").

The parts are subjected for about 5 minutes to the fumes of equal volumes of concentrated nitric and hydrochloric acids, then heated to 300°C. (abt. 600°F.) until the bronze colour appears. After cooling the pieces are greased with vaseline, heated until this has evaporated, then finally oiled after cooling.

The coating obtained has a reddish-brown shade: an addition of acetic acid produces a bronze-yellow, and by different proportions of the mixture shades from pale to dark reddish- and yellow-brown can be obtained.

When carefully carried out the rust-protection afforded is considerable, and iron parts so treated resisted the acid fumes of a chemical laboratory for 10 months.

> *Note.* Of little value for gun purposes, this process has only been quoted here for completeness and use on occasion for other purposes.
>
> The mixture of chlorine and nitrous fumes is extremely corrosive and deleterious, and attacks metal violently, so that this treatment is best carried out in an open shed or specially well ventilated room.

K. Browning "Antinit" and "Anticorro" Steel

These special barrel metals, made respectively by the firms of Gebr. Böhler, Kapfenberg, Austria, and the Poldihütte, Kladno, Czechoslovakia, are extensively used by continental European (and to a lesser extent by American) gunmakers for luxury guns. They are hard, extremely tough nickel steels, and although not "rustless" to the extent of so-called "stainless" (high-chromium) steel, they resist wear and corrosion by damp and residues of charge combustion to a remarkable degree, a great boon to the lover of fine arms.

The author's notes on previous experience with these metals having gone astray in the course of removals, the undermentioned results were obtained with trial pieces, remarking that these experiments took place during a period of hot, dry weather which rendered rusting abnormally slow, even allowing for the inherent rust-resistance of the pieces.

The slow action and unsatisfactory results given by certain brownes of particular excellence for plain carbon steels suggested repetition of the trials, first "priming" the pieces with 2% ammonium chloride solution, and reapplying "primer" and browne alternately twice at each repetition, boiling for 20 minutes between each as usual. The brownes used were those immediately available, but it is of course not suggested that others may not be as good or better. All rustings took place at room temperature. The references are the Group numbers of the brownes applied.

Böhler-"Antinit": unprimed brownes reapplied after drying.

Aa. 22, Ac. 1. Alone: unsatisfactory brownish tone after 3 repetitions. Primed: splendid deep black after 2 repetitions, equal to that obtained with these brownes on plain carbon steel. Slight after-rusting, neutralisation needed.

B. 5, B. 12. Alone: very good deep black after 3 repetitions. B. 12 slower than B.5, but final result equally good.

C. 13. m. (Swiss Federal Armoury browne, normal composition) and **C. 13. n.** (fuming instead of plain nitric acid, no ethyl nitrite or alcohol). Both alone: 3 repetitions gave an equally fine, deep black, C. 13.m. acting very slowly, C. 13.n. sensibly quicker, but slightly after-rusting: neutralise.

C. 19 ("Ferrobronze"). Alone: 3 repetitions gave a very fine, deep black. Quicker than D. 11.t.

D.11.k, (m) and (t). Alone: (k) extremely slow, but good deep black with 4 repetitions: (m) and (t) successively quicker, and equally good result with 3 repetitions.

D. 19. c. ("Ferrooxydin", salt-prepared). Alone: quicker than C. 19, and excellent deep black with 2 repetitions.

Poldi-"Anticorro": in general, rusts more slowly than "Antinit" with the same brownes as above, and same procedure.

Aa. 22, Ac. 1. Alone: slow, but good deep black after 4 repetitions. Primed: very fine result with 2 repetitions, equal to finish obtained on plain carbon steels. Very slightly after-rusting, but neutralisation advisable.

Ab. 6. Alone: 4 repetitions gave a very good, deep black.

B. 5. Primed: very fine black with 3 repetitions.

C. 13 (Swiss Federal Armoury browne). Alone: brownish tone after 4 repetitions, "normal" composition (m) acting very slowly, (n) rather more quickly. Primed: 3 repetitions gave an excellent deep black.

C. 19 ("Ferrobronze"). Alone: 4 repetitions gave an unsatisfactory brownish tone. Primed: excellent deep black with 3 repetitions.

D. 11.m. Primed: fine deep black with 3 repetitions: slower than B.5.

D. 19.c. ("Ferrooxydin", salt-prepared). Alone: brownish tone after 4 repetitions. Primed: excellent deep black, as C. 19, with 3 repetitions, but quicker than C. 19.

CHAPTER V

PICKLING, ETCHING & MATTING

Pickling (Fr. "décapage," G. "Beizen") is an occasional preliminary to browning, and is applied to the following purposes:

A. Scaling (Fr. "dérochage," G. Blankbeizen").
B. Priming (Fr. "amorçage," G. "Einleitung") action of browne.
C. Matting (Fr. "matage," "mise au mat," G. "Mattieren").
D. Etching (Fr. "décapage," G. "Anätzen") twist barrels to bring out figure.

These applications partly shade into each other: pickling proper (scaling) is intended to remove oxide, in order to provide a clean metallic surface for further operations, the gentler "etching" to dull (Fr. dépolir," G. "abtönen") already highly polished surfaces, or to bring out the beautiful wavy figure of twist barrels.

Pre- and after-treatment of parts are summarised at the end of this section.

A. Scaling

1. Plain acid bath.

Oxide is easily removed by a short immersion in weak (5 to 10%) sulphuric, nitric or even acetic acid, by which it is loosened and can then be readily wiped off, leaving a clean, dark grey metallic surface, from the carbon contained in the metal.

2. Bright pickle. ("Metallarbeiter").

Sulphuric acid D.1.84	82 grs.	4.5 g
Nitric acid D.1.42	37 "	2.0 "
Zinc	7 "	0.4 "
Water t.s. to make	¼ U.S. pt.	100.0 ccm

First dilute the sulphuric acid in a portion of the water, add the nitric acid, the rest of the water and the zinc, in this order, and do not use until the mixture has completely cooled. This pickle leaves a bright surface, thereby saving labour in repolishing.

130

3.   Quick-acting bright pickle (F. W. Würker, Brit. P.
260,456/26).

Pickling is considerably accelerated by the addition of small
quantities of organic substances containing at least one alcohol
radical, to the mineral acids.

As an example, the addition of 5 to 8 g acetone and 10 to
20 g of butyl alcohol to each liter of 10% sulphuric acid (73,
116, 146 and 300 grs. per U. S. quart) is recommended. Such
a pickle is stated to act only on the layer of oxide, not on the
metal, so that it not only economises acid but also leaves a clean
metallic surface.

4.   Bright pickle for rustless steel. (J. Sankey & Sons Ltd.,
Brit. P. 225,416/24).

Chlorine-evolving mixtures, e. g. aqua regia, mixtures of hydrochloric acid and manganese dioxide, chloride of lime and
sulphuric acid, etc. are specially suitable for cleaning and matting
rustless steel. A 30% hydrochloric acid, warmed to about 50°C.
(120°F.) is recommended, to which the manganese dioxide is
added in successive small portions: after pickling, neutralise with
hot or cold caustic lye.

> *Note.* On account of the extremely deleterious nature of chlorine,
> this process is the reverse of pleasant, and should only be
> undertaken in the open or in a specially well ventilated room.

Old browning has occasionally to be cleaned off for reconditioning, which can be effected by the use of one of the pickles
described (and many others): but the employment of strong
mineral acids is always attended by risk, as there is no positive guarantee against carelessness, accidental or other. Being
essentially a derusting operation, one or other of the methods
described in Chapter VII under "Derusting" is much to be
preferred, and in particular, one not including mineral acid
(except the very mild boric acid), which attain the desired end
with thoroughness and without trouble.

B. "Priming" action of Browne

As repeatedly mentioned, a slight etching of the pieces to be
browned in many cases facilitates the initiation of action and
often improves most markedly the appearance of the final finish.
Herewith a few suitable etchants, to which in certain cases the
methods described under "Derusting" can be added with
advantage.

1. Salammoniac priming.

This has already been mentioned in Chapter II. 4, under "Rusting," paragraph "Casehardened parts," to which the operator is referred: the "kink" is most simple and effective, and the parts "primed" require no further treatment before applying the browne. Used cold.

2. Plain acid etchant.

Nitric acid in 5% strength produces a slighter, in 10% concentration a deeper etching of the surfaces suitable for gun sights and similar parts. For barrels and larger parts, use as described below.

3. "Jewellers' etchant."

Plain nitric acid has little or no action on certain steels, such as Böhler-Antinit, Poldi-Anticorro or stainless properly speaking, and with these the so-called "jewellers' etchant" (known in the U. S. as "Spencer acid") renders useful service. It consists of

Silver nitrate	55 grs.	3.0 g
Mercurous nitrate	55 "	3.0 "
Nitric acid D.1.42	55 "	3.0 "
Water t.s. to make	¼ U.S. pt.	100.0 ccm

On account of the sensitiveness of the silver salt to light, this etchant should be kept in a brown glass stoppered bottle and in the dark: in itself it is quite stable. (Avoid confusing mercur*ous* nitrate, L. hydrargyrum nitricum oxyd*ulatum* with the mercur*ic* salt, L. hydrargyrum nitricum oxyd*atum:* Fr. "nitrate mercur*eux*," G. "salpetersaures Quecksilberoxy*dul*").

Application. Both plain nitric and jewellers' etchant are used in the same way. The parts to be etched are laid in boiling water until they have taken its temperature, then wiped over as quickly and evenly as possible with a pad of cotton wool dipped in the etchant, emphasizing speed and uniformity. The etchant dries quickly on the hot metal, its surface assuming a silky, silvery appearance: if any spots remain bright, warm again in boiling water and repeat.

4. Bright etchant for non-rusting steel (J. H. G. Monypenny).

A very fine silvery surface can be obtained on non-rusting steel by a cold dip containing from 3 to 5% of nitric and 1% of hydrochloric acid (weak aqua regia). Its action is rather slow, and the etchant is specially suitable for removing thin layers of oxide, to which end a period of ½ to 1 hour is usually sufficient, although heavier coatings can be removed by prolonging immersion to as much as 24 hours if need be.

C. MATTING

This operation (Fr. "matage," "mise au mat," G. "Mattieren") is partly covered by the preceding, but is more correctly a stage in finishing proper. It imparts a velvety, very refined appearance to the browned parts, combined with the very practical advantage of a non-reflecting surface, eliminating the light-reflections which dazzle the shooter and may reveal his position to game or enemy. Depending on conditions and nature of the surfaces to be treated, matting can be carried out by a variety of mechanical or chemical means.

Mechanical matting includes guilloching, graining with sand-blast and matt brushing, and for smaller parts stippling and spark-spotting.

Guilloching, (Fr. "guillochage," G. "Guillochieren") also called "engine-turning," is a purely mechanical operation often applied to the barrels and receivers of luxury arms, and is practically confined to these parts. Its beautiful wavy effect is highly decorative and is indeed its chief recommendation, as for purely mechanical reasons it cannot be extended to the entire surface, and consequently does not completely annul reflections. Greatly esteemed for its beauty, it does nevertheless greatly diminish this evil and has a positive, practical shooting-technical value.

The closely related, purely mechanical operation of *knurling* is practically restricted to round parts, and is less effective, as the summits of the small pyramids produced by the knurl soon wear bright by friction.   (The barrel of the Winchester automatic shotgun had a knurled belt of very pleasing appearance, whose purpose was to afford handhold).

Sandblasting (Fr. "passage au jet de sable," G. "Sand-strahlbehandlung") meets all requirements of the gunmaker and innumerable others, and is indeed a process of rare universality of application, extraordinarily effective and flexible: according to the matting medium used—pumice, glass powder, chilled iron shot, "steel grit," quartz sand or emery of different grain—and air or steam pressure, it can produce any desired effect at will, from the most delicate shadow matting to complete scaling of the roughest castings or ingots. In addition, the treated surfaces are absolutely clean and grease-free, and after brushing, preferably with a circular brush, to get rid of adhering loose dust, the pieces can immediately be passed on for browning.

Motive power is a necessity, and the sandblast is an appliance of enormous capacity, so that it is really only suitable for operation on an active manufacturing scale: for general gunsmith's purposes recourse must be had to outside help. Its only disadvantage is the production of an enormous cloud of the finest dust, which, unless used in the open, renders a dust-collecting arrangement practically indispensable, which again needs power and adds to the expense of the plant.

Matt-brushing ("frosting," Fr. "dépolissage," G. "Mattschlagen"). This process is only suitable for light matting of comparatively soft material, as this must obviously not be harder than the wires of the brush. Matting brushes differ from scratch brushes in having specially long wires, not tightly bound into a centre, but assembled in tufts loosely hinged or ringed to the centre. They are driven at much higher speeds than scratch brushes (up to 90 ft./sec. at the periphery) and the work-pieces are presented to the brush so that the points of the wires impinge like needles on their surface, instead of stroking as in scratching (this would produce a bright finish). On account of the small difference in hardness between the wires and work-pieces when used on steel, the matting effect is comparatively mild, and its grain roughly proportional to the thickness of the wire. The "velvet finish" adopted by Smith & Wesson Inc. during the Great War was obtained by this method, and for beauty and practical value left nothing to be desired.

Stippling (Fr. "pointillage," G. "Punktieren," "Tüpfeln"). Smaller parts of not over hard material can be stippled by rapid blows with a sharp centre punch (Fr. "pointeau," G. "Mattpunze," "Körner"), but much skill and patience are requisite to produce a pleasing effect. The process is slow, but can be accelerated by using a multi-pointed punch or even a small piece of a rather coarse file, but usually with unpleasing effect, as it is next to impossible to align the groups of dents with absolute regularity.

For the fairly busy gunsmith (or well-to-do amateur) a very much better and speedier method is to use a "dental engine," preferably with electric drive, the convenience and many-sided applicability of which to a large variety of small mechanical jobs has rightly been emphasized by C. Baker ("Modern Gunsmithing"), and is such as to outweigh its cost in a very short time. "Burrs" (dental drills) in a very large variety of shapes, worn too dull for dental purposes, but keen enough to be extremely useful to the gunsmith, can be had from any dentist almost for the asking, and the variety of special attachments to the engine enable operations to be carried out with ease that would otherwise be next to impossible. Using a sharp punch held in the percussion attachment (used in "stopping"), parts can be stippled with ease and far greater rapidity and infinitely better appearance, owing to the uniformity of the impacts, than can possibly be done by hand. Casehardened parts, if not excessively hard, can be dealt with by a punch of high-speed steel.

After stippling, the burrs raised by the punch can be removed by the careful use of a smooth file, oilstone, emery cloth, etc.

Spark-spotting (Fr. "pointillage électrique," G. Funkbrennen"). Stippling can be done just as rapidly, if with less pleasing effect, by the electric spark, requiring only the simplest equipment, a 6- to 12-volt accumulator (e. g. storage) battery, and an electrode consisting of a pointed copper rod held in an insulating handle. One pole of the battery is connected to the work-piece, the other to the electrode (by ordinary flex) ; rapid making and breaking contact between electrode and work, the sparks of rupture burn slight dents in the metal, and all that is needed is to go systematically over the entire surface to be stippled. The dents are naturally shallower than those produced by a punch, but the general effect is quite acceptable: a particular advantage of the process is that parts of extreme hardness, which could not be stippled by any other means, are as easily dealt with as soft ones.

Where alternating current is available, a small transformer of at least 50 watts capacity, advantageously replaces the battery. (The spotter made by the Arkograph Co. of Portland, Ore., is a cheap and effective appliance on these lines: more elaborate is the "Etchograph" (used at Springfield Armoury for marking rifle bolts), the "Signierapparat" of the A.E.G. Co., Berlin, and similar appliances made by the large electric companies.

Chemical matting. This method requires the absolute minimum of work and gives quite pleasing effects, even if not quite equal to those produced by sandblasting.

For matting purposes the treatment described in Chapter VI for derusting with potassium bisulphate or boric acid is specially suited. According to concentration and time of action, the former produces a well-defined matt, the less soluble and milder-acting boric acid a very fine, delicate dulling of the polished surfaces, with the minimum of trouble and with every satisfaction in result.

For the conditionally applicable but not entirely irreproachable method by after-rusting, see Chapter II. 7. "Neutralising."

D. Etching twist Barrels

The methods here described are based on the differential action of the chemicals used on the iron and steel strips of the twist.

The oldest, purely mechanical method of bringing out the figure of the twist, recommended by F. Brandeis for black browned (boiled) barrels, is described in Chapter II. 11. Considerable skill and practice, as well as patience, are needful to produce a pleasing effect, and the method is relatively slow and expensive.

Chemical methods attain the end with ease and certainty, and a few are here described.

1. Steele & Harrison. According to the effect desired, the unbrowned barrels are laid for 15 to 30 minutes in saturated copper sulphate solution, in which they become covered with a pink deposit of metallic copper. This is rubbed off with a bristle brush or oily rags, rendering the wavy figure apparent. Repeat if necessary, then proceed to brown.

2. According to **Buchner** the barrels are laid for 3 to 4 hours in quite weak hydrochloric acid (maximum 1%, that is, 30 g D.1.16 per liter = 140 grs. per U. S. quart) then rinsed, rubbed off with tow and tripoli, dried, oiled and warmed over charcoal fire.

3. F. Brandeis states that this method is not particularly good, and that rinsing with boiling 15% sulphuric acid is to be preferred, then thorough washing and drying. Very weak nitric acid can also be used, but this often leaves the barrels darker after etching: it is best to warm the barrels in hot water before laying them in the etching trough.

4. Another etchant recommended by **F. Brandeis** is composed of

Nitric acid D.1.42	28 grs.	2 g
Acetic acid	1450 "	100 "
Ammonium chloride	14 "	1 "
Water t.s. to make	1 U.S. qt.	1 liter

to which a little copper sulphate should be added to prevent over-energetic action of the nitric acid, and to produce a more even effect: the fine copper deposit is easily rubbed off with an oily rag. This etchant is slow-acting, occasionally even without the copper sulphate addition.

5. Other cold etchant by **Brandeis.**

Mercuric chloride	730 grs.	50 g
Nitric acid D.1.42	59 "	4 "
Tartaric acid	460 "	32 "
Water t.s. to make	1 U.S. qt.	1 liter

The barrels are etched in two stages, duration according to distinctness of the figure. Following the usual after-treatment, the barrels are brushed with the scratch brush and fine pumice powder, causing the harder strips to appear lighter, the softer ones darker.

> *Note.* The above etchants embody the use of strong mineral acids, which always offer the possibilities of mischief from inattention. The author would suggest potassium bisulphate solution as a milder alternative (at bottom this is equivalent to diluted sulphuric acid) as milder and more easily controlled, but having had little experience with twist barrels, merely offers this as a suggestion.

Pre-and after-treatment of barrels. All parts to be etched or matted must first be thoroughly degreased, barrels and other hollow parts tightly plugged, and surfaces not to be matted protected by a suitable varnish.

After etching, all acid and salt residues must be most thoroughly removed: to this end, repeated rinsing with boiling water, then with a little added potash or soda, and again with clean boiling water, drying, occasionally over charcoal brazier, finally oiling, are needed.

CHAPTER VI

DERUSTING AND DENICKELLING

When reconditioning guns or pistols, parts more or less badly rusted, or with partly worn-off browning or nickel plating, often have to be dealt with. Smaller, not too deeply rusted parts of simple form can at a pinch be freed from rust mechanically, e. g. with emery cloth, larger ones by a coarser wire brush (thickness 8 to 10/1000", peripheral speed 65 to 90 ft./sec.), but the mechanical derusting of barrels or parts of complicated shape, in the absence of a tumbling barrel or bowl, is always slow and tiresome.

Chemical means attain the desired end, regardless of shape and thickness of rust layer, with the minimum of trouble, and complete thoroughness and certainty.

As with browning, thorough degreasing is an indispensable preliminary to derusting and denickelling.

A. Derusting

Derusting (Fr. "dérochage," "dérouillage," G. "Entrosten") is simply accomplished by laying the rusted parts in a solution until the rust has entirely disappeared: nothing could be simpler. The following are some of the better-known and more recommendable methods.

1. Oxalate derustant. This is one of the best known methods: a 3 to 5% solution of acid potassium oxalate (L. kalium bioxalicum, popularly salts of sorrel or of lemon, Fr. "sel d'oseille," G. "Kleesalz") dissolves the rust, slowly at room temperature, but fairly rapidly in gentle warmth (50°C. = 120°F.).

2. Oxalic acid. According to the author's experience, a 2 to 3% solution of oxalic acid (L. acidum oxalicum) is equally satisfactory, appropriate after-treatment assumed.

3. Potassium bisulphate. The parts are laid in a 3 to 5% (at most) solution of this salt (L. kalium bisulfuricum, acid potassium sulphate) and touched with a piece of zinc to initiate action, upon which an evolution of hydrogen immediately commences. Derusting is fairly rapid at room temperature. Note that this solution is equivalent to diluted sulphuric acid: a higher concentration is without object and may indeed heavily corrode the parts.

The solution keeps indefinitely, so that it can be held in stock and used until completely exhausted. Old browning is quickly and thoroughly cleaned off: as noticed under "Matting," this solution affords a very convenient way of matting parts for which a dull surface is desirable.

4. Boric acid. This is a specially excellent, mild matting and thorough derusting agent. According to the author's experience, a 6 to 10% solution is very suitable for derusting: on account of the small solubility of the acid, gentle warming of the bath (30° to 50°C. = 85-120°F.) is recommended, at which temperatures the concentrations given are nearly those of saturation. As a matting agent, boric acid is specially recommendable for light matting, leaving a very fine, delicate dull surface on finer parts.

Its action is slower than that of most derustants (boric is a weak acid at ordinary temperature) but it thoroughly clears away the thickest rust and only affects the cleaned surfaces to a minimum extent: in addition to this, the derusted parts are less inclined to after-rust than with most other derustants.

5. Acid reduction. According to Dr. Kropfhammer, D.R.P. 375,426/22 rust is reduced and loosened by less corrosive acids, such as phosphorous, hypophosphorous and formic acids, alone or mixed.

A 10% solution of phosphor*ous* acid (L. acidum phosphor-*osum*, not to be confounded with phosphoric acid, L. acidum phos-phor*icum*) or formic acid (L. acidum formicicum) can for example be used. The derusted surfaces remain black, contrary to their state after using other derustants, but are easily rubbed bright.

6. Cyanide derustant. According to F. Voytier (and others) the rusted parts are moistened with a 50% solution of potassium cyanide, then rubbed with a paste of 1 part of this salt, 1 of fatty soap and 2 of precipitated chalk, well mixed together.   After a little time the rust softens and can gradually be rubbed off.   Polished surfaces are unaffected.

> *Note.* Apart from the relatively complicated procedure, the utmost care must be taken in handling this chemical, one of the most violent poisons known: accidental contact with an acid evolves deadly fumes (of prussic acid gas) and even a slight wound on the hands may have fatal effect.

7. Sodium sulphide. In D.R.P. 394,819/23 F. von Karlowski describes a process based on conversion of the rust into ferrous sulphide.

A 5% sodium sulphide solution for instance converts the layer of rust, according to its thickness, into sulphide in from 4 to 36 hours; frequent shaking accelerates action. The parts are then rinsed and laid in 4% hydrochloric acid, which decomposes the sulphide in 2 to 12 hours, depending on its thickness, or in less time if frequently stirred.

> *Note.* This "2-stage" process is quoted for completeness, but is of little advantage to the gunsmith.   In the first place, other methods are simpler, secondly, the sulphuretted hydrogen given off in stage 2 is the reverse of agreeable (smell of rotten eggs) and finally, the use of strong mineral acid is best avoided at all times.

8. Zinc derustant. As is well known, zinc is dissolved by caustic alkalies with evolution of hydrogen, on which action the following derusting method is based.

To 100 ccm of 20% caustic soda lye about 15 g (365 grs. in ¼ U.S. pt.) of zinc dust, which usually contains from 5 to 10% of zinc oxide are added. At room temperature no action takes place, but when warmed to 60° to 80°C. (140-175°F.) a lively evolution of hydrogen sets in, which completely reduces the rust, according to its thickness, in from 2 to 3 hours.

The hydrogen evolved is inodorous, and as derusting takes place in alkaline solution, no further after-treatment to prevent re-rusting than plain washing with hot water is required. Polished parts of the surface are quite unaffected. The solution is best made up fresh as required.

9. Tin oxalate (L. stannum oxalicum) in solution acidified with hydrochloric acid can likewise be used as a derustant, for which purpose Frémy states that a 4 to 6% concentration is sufficient.

10. Tin salt. The reducing action of this salt (stann*ous* chloride, not to be confused with stann*ic* chloride) on iron and other oxides is well known and largely used in technical industry. A suitable solution is given by Frémy, and consists of

Stannous chloride cryst.	880 grs.	60 g
Hydrochloric acid D.1.16	95 "	6.5 "
Water t.s. to make	1 U.S. qt.	1 liter

the action of which at a temperature of 50° to 60°C. (120-140°F.) is rapid and complete. Hydrochloric acid is necessary to prevent precipitation of basic salt.

11. The process described by Dr. Aug. Buecher in D.R.P. 52,162/89 is based on the action of stannic chloride (tin tetrachloride) : of which he described two working methods.

A. Starting point is ready-made stannic chloride. The bath consists of

Stannic chloride anh.	730 grs.	50 g
Tartaric acid	18 "	1.25 "
Sulphuric indigo solution diluted		
100 times	9.5 ccm	10 ccm
Water t.s. to make	1 U.S.qt.	1 liter

The chloride is first dissolved in about half the water, generating great heat: after cooling, the previously dissolved tartaric acid and indigo solution are added and the whole brought to volume. (The strongly fuming, irritant liquid anhydrous chloride can equally well be replaced by its equivalent 67 g (980 grs. per U.S. quart) of hydrated chloride or butter of tin).

This solution contains no poisonous substance, so that table knives and surgical instruments can be derusted by it.

B. Stannous chloride (tin salt) and mercuric chloride are the starting point, by which stannic chloride is generated in the bath. The proportion of the different ingredients varies according to the thickness and density of the rust to be removed: if these be not great, the following composition is very suitable, even for very delicate parts.

Stannous chloride cryst.	145 grs.	10 g
Mercuric chloride	29 "	2 "
Tartaric acid	44 "	3 "
Sulphuric indigo solution diluted		
100 times	48 ccm	50 ccm
Water t.s. to make	1 U.S. qt.	1 liter

Mercuric chloride being a violent poison, this solution can only be used when this is of no disadvantage.

The parts to be derusted are left in one or other of these solutions for 24 hours, or as long as necessary: the rust is completely eliminated, leaving a clean metallic surface, and the polish of bright surfaces is not affected. After-treatment must follow *immediately* after derusting.

> *Note.* Solution (A) is very stable, and the author found un-diminished activity in some that had been kept for 7½ years in a stoppered bottle. Solution (B) will not keep indefinitely, but after long standing becomes cloudy and deposits a gelatinous mass consisting mainly of metastannic acid.

> *General remarks on derusting.* It will have been noticed that the effect of most of the derustants is based on the action of nascent ("in the act of liberation") hydrogen, which energetically reduces and thus breaks up the oxides.

> Although all the derustants here described make a clean sweep of the rust, those which include powerful mineral acids are preferably avoided, as the derusted parts quickly become covered with a film of fresh rust: this relapse is favoured by mineral acid, and also by mercuric chloride, an unfortunate accompaniment of the otherwise excellent stannic chloride derustant (B). Speedy and thorough after-treatment is the only way of avoiding this unpleasing action, and in this respect the zinc-soda derustant is the best, closely followed by boric acid.

After-treatment. On removal from the derustant the parts are covered with adherent liquid and loose, more or less dark grey powder; now and then black spots remain, due to the carbon content of the metal, but which are of no importance and are easily rubbed off.

To prevent re-rusting, the parts should *immediately* be dropped into hot, preferably boiling water, to which a little soda or potash can advantageously be added, and brushed with a stiff bristle, or if necessary, brass wire brush. The water should be renewed once or twice, finally washing with clean water and drying as rapidly as possible.

The metallic surfaces are then quite clean, but the metal eaten away by rust is irremediably gone, so that the piece must be repolished all over if its appearance is to be at all satisfactory: but scars left by deep rusting are incurable. If repolishing is not to be done immediately the parts should be oiled, but if browning is to be undertaken next, but only after a certain interval, they should be kept (if possible) in a completely dry spot, in a desiccator (if small), or laid in boiled, that is, gas-free water to which from 5 to 10 grams per liter (72 to 145 grs. per U.S. qt.) of cream of tartar or slaked lime be added.

All in all, the author has for choice chiefly used the stannic chloride (A), zinc and boric acid derustants, now and again bisulphate for less particular parts, with complete satisfaction.

B. Denickelling

(Fr. "dénickelage," G. "Entnickeln"). As for derusting, chemical action offers the readiest and least troublesome means of denickelling parts on which the nickel plating has become partly worn off, with the advantage over electrolytic action that damage to the underlying iron or steel is positively excluded.

The most practical method is that developed by the King's Norton Metal Co. Ltd., Birmingham, England, described in Brit. P. 13,298/05 for denickelling rifle barrels, and based on the action of ammonium persulphate, which in presence of ammonium carbonate and free ammonia dissolves copper and nickel, but does not attack iron.

The so-called "K.N.S." solution consists of:

Ammonium persulphate	5 g
Ammonium carbonate	1 "
dissolved in 100 ccm of ammonia D.0.935.	

These proportions are often varied, and an addition of a little ammonium dichromate is occasionally recommended, but of no real advantage. In the U.S. **a stronger solution** is usual, its composition (in round figures) being

Ammonium persulphate	10 g
Ammonium carbonate	5 "
dissolved as before in 100 ccm of ammonia D.0.935.	

The solution is in itself quite stable and will keep indefinitely in glass-stoppered bottles, and does not attack iron so long as loss of ammonia is kept to a minimum. Therefore keep away from direct sunlight and in as cool a place as possible, and leave bottle open for the least possible time.

The degreased parts are laid in the cold solution, preferably in a wide-mouthed, glass-stoppered jar, preserving jar with air-tight cover or other vessel that can be securely closed. If copper is present the solution quickly becomes blue: denickelling is complete when the uniform grey colour of the underlying iron appears on the whole surface of the piece. Then rinse, dry and oil or brown as required.

> *Note.* Ammonia density 0.935 is obtained by diluting ammonia D.0.880 with an equal volume of water. If this is not available, the desired concentration is also obtainable by diluting 63 ccm of "stronger ammonia" Ph.U.S.A. or 55 ccm "stronger ammonia" Ph.Brit. to 100 ccm, using only distilled water for this purpose.

As an indication of the effectiveness of this method, the English (weaker) K.N.S. solution removes most of the cupro-nickel casing of a German M/88 7.9 mm bullet (made by the D.W.M., Karlsruhe) in about ¾ hour, and the thicker layer of plating at the point in about 2 hours.

With 100 ccm of the English composition from 6 to 7, and of American composition (as above) from 12 to 14 such bullets can be completely cleared of cupronickel.

Nickel or nickelled parts can moreover be easily and very neatly matted by suitably timed immersion in this solution.

CHAPTER VII

PHOSPHATISING

A few notes on this process are here added for completeness:
the process is not intended for metal colouring, but for obtaining
a coating of greater rust-protecting quality than is possible by
browning or other treatment, which by its intimate union with
the underlying metal will resist damp where other coatings fail.

Phosphatising is a collective process equally applicable to
small or large scale operation, for products varying between
cycle and automobile accessories to entire frames and even
heavier pieces for military and naval use, now as a cheap mass-
produced finish, again as a grounding for enamelling, for which
latter purpose a non-rusting base is essential to prevent peeling.

It is only occasionally used for parts of small arms, e.g. for
the receivers of Springfield rifles: according to their catalogue,
the Newton Arms Co., of Buffalo, N. Y., finished their barrels
by this process.

Phosphatising is less suitable for fine work, as the principle
postulates an attack on the metal surfaces exceeding the limits
of standard matting, so that phosphatised parts mostly have a
dull, when not positively rough surface: polishing with a circular mop and oil somewhat improves the appearance, but it
is always inferior to that of a finished and browned or blued
part. For parts like front and rear-sights it is however of positive advantage, and to this extent the process is of interest to
the gunsmith.

Being a patented process, precise working instructions are

145

only given to licensees: so far as they are contained in the patent specifications they are reproduced here.

The essence of the phosphatising process is prolonged boiling of the work-pieces in diluted phosphoric (orthophosphoric) acid, the earliest record of which is contained in Brit. P. 3119/69 of Colonel W. A. Ross, which however gives only vague indications of its mode of application: the much later developments of T. W. Coslett, Birmingham, and Richards, Coventry, England, first gave it practical value. In England it is generally known as "Coslettising" or "Fermanganising," in the U.S. as "Parkerising," from the name of the exploiting firm in Detroit. The earlier patents are time-expired and can be used by anyone.

1. T. W. Coslett, Brit. P. 8667/06, D.R.P. 209,805/07, U.S.P 870,937/07.

The working bath is composed of

Phosphoric acid D. 1.5	365 grs.	25.0 g
Iron filings or shavings	91 "	6.25 "
Water t.s. to make	1 U.S. qt.	1 liter

The degreased parts are boiled until the bath has evaporated to a fraction, e.g. 1/7 of its original volume, which may last from ½ to 3 hours, and produces a greenish-grey-black coating, somewhat darkened by oiling. The metal is slightly attacked at the beginning of the process, but this action ceases with the formation of the coating, which consists essentially of basic phosphate.

Prior sandblasting is stated to improve noticeably the appearance of the phosphatised pieces, whose surface is naturally dull.

2. Anti-rust Syndicate and Coslett, Brit.P. 22,743/09.

According to this patent a concentrate is first prepared with iron filings and phosphoric acid, which when diluted 80 times forms the working bath, containing

Phosphoric acid D. 1.5	91 grs.	6.25 g
Iron filings	37 "	2.5 "
Water t.s. to make	1 U.S. qt.	1 liter

in which the parts are boiled for 1½ to 2½ hours, with the same result as before. This modification avoids the excessive formation of suspended iron phosphate, which tends to settle in hollows, slots, etc. and the extensive dilution of the bath avoids injury to polished surfaces. The patent covers the revival of the bath by periodical decantation and addition of concentrate.

3. F. R. G. Richards, Brit.P. 17,563/11, D.R.P. 265,-249/12, U.S.P. 1,069,903/13.

This modification, known as the "Fermangan" (Fr. "fer-manganèse") is characterised by the addition of manganese dioxide to the bath: this consists of:

Phosphoric acid D. 1.5	61 grs.	4.2 g
Manganese dioxide	37 "	2.5 "
Water t.s. to make	1 U.S. qt.	1 liter

The pieces are boiled for ½ to 1½ hours, either making up the water evaporated, or evaporating it to 1/3 of its original volume. As the manganese dioxide does not dissolve but remains in suspension, the bath must be kept boiling briskly to secure a uniform coating. Further evaporation results in rougher coatings covered with fairly strongly adherent powdery manganese dioxide.

The coating consists of mixed basic manganese and iron phosphates, is rougher and noticeably darker than the Coslett-produced and offers considerably greater resistance to rust.

4. Parkerising is described by Field & Bonney as follows, the bath consisting of

Phosphoric acid	365 grs.	25.0 g
Manganese dioxide	22 "	1.5 "
Water t.s. to make	1 U.S. qt.	1 liter

Cast-iron filings can be substituted for the manganese dioxide. The pieces are boiled in this for 2 to 4 hours, then rinsed first in cold, then in hot water, dried in sawdust and brushed to free them from adherent sawdust. They are then heated on a hot plate to "hissing" temperature, at which water dropped onto it is immediately thrown off, and at this temperature dipped in linseed oil, drained and gently warmed until dry. This is simply the Coslett or Fermangan treatment.

The above data are sufficient for the occasional phosphatising of smaller parts, for which finished appearance is of secondary importance, and where special measures to promote economy would be unprofitable on account of the very small quanities of ingredients used.

Industrial application however postulates economical working: in the course of time various troubles have manifested themselves, and a variety of patents have been taken out with the object of overcoming these and of securing regular and profitable results.

After-treatment. The finished parts are transferred without rinsing to a drying oven or hot sawdust, so as to dry as quickly as possible, then dipped in linseed or vaseline oil, which sensibly darkens the colour. Their appearance is improved by rubbing or brushing with a bristle brush and a mixture of 1 part solid paraffin and 2 parts turpentine, or one of 3 parts mineral oil and 1 part turpentine, which also slightly increases their resistance to rust.

According to the very thorough investigations of the U. S. Bureau of Standards (Nr. 80), the rust protection secured by phosphatising is not absolute: it gives a certain degree of protection to moderate exposure and mechanical action: frequent oiling increases its resistance to weather, but not to mechanical influences.

On casehardened parts it fails unless the hard skin is first in greater part or entirely removed by sandblasting or pickling.

APPENDIX

AMERICAN, BRITISH AND METRIC WEIGHTS AND FLUID MEASURES

American and British weights, so far as concerns the subject in hand, are identical, and the system (or muddle) of Avoirdupois, Troy and Apothecaries' weight persists in both lands. Occasionally all three are to be found in a single browning formula, to the confusion of correct interpretation, but in the absence of clear evidence to the contrary, avoirdupois weight has been assumed throughout the present work. The "grain" is the only unit common to all systems, for which reason it has been exclusively used in "English" formulae in the present study. The difference between "dram" and "drachm"—unfortunately also pronounced "dram"—should be specially noted, as a mistake e. g. in weighing a powder charge would have unpleasant consequences.

The differences between American and British fluid ounces and pints, etc., have already been dealt with in Chapter IV, likewise the value of the "drop" according to International Convention.

Metric units have the advantage of always meaning the same thing, and of interconnection in such a way as to present the simplest possible relation between weight and volume.

Older German formulae are occasionally to be found expressed in the old "Nuremberg Medicinal and Apothecaries" weights, for which reason their metric equivalents are likewise given here.

In the following table, Av. means avoirdupois, Tr. troy, and Ap. Apothecaries' weight.

Weights (U. S. A. and Britain).

Unit.	System.	Value, grams.
1 grain	All.	0.0648
1 scruple = 20 grains	Ap.	1.296
1 pennyweight (dwt.) = 24 grains	Tr.	1.555
1 dram = 27.344 grains	Av.	1.772
1 drachm = 3 scruples = 60 grains	Ap.	3.888
1 ounce = 16 drams = 437.5 grains	Av.	28.349
1 ounce = 24 scruples = 20 dwts.	Ap. & Tr.	31.104
1 pound = 12 ounces Tr.	Ap. & Tr.	373.241
1 pound = 16 ounces Av.	Av.	453.593

Fluid Measures.

Unit.	U. S. A.		Britain.	
1 minim (drop)	0.0616	ccm	0.0592	ccm
1 fluid drachm=60 minims	3.697	"	3.552	"
1 fluid ounce=8 fl. drachms	29.573	"	28.417	"
1 pint (a)	473.168	"	568.336	"
1 quart=2 pints	946.34	"	1136.67	"

(a). 16 fluid ounces in U. S. A., 20 do. in Britain.

When it cannot be definitely established whether a browning formula in English is of British or American origin, a "compromise" fluid ounce of 29 ccm can be assumed, which is almost the exact mean of the American and British units. The corresponding concentrations are then about 4% under (U. S.) or over (Brit.) the correct value, were the origin of the formula known.

1 gram   =15.432 grs.   100 ccm= 3.38 U. S. or 3.52 Brit. fl. oz.
100   "   3.53 oz. av.   1000   "   2.11 U. S.   1.76 Brit. pt.
                              (1 liter)
1000 " (1 kilogram) 2.2046 lbs. av.

Old Nuremberg Weights.

1 Gran (grain)	0.0609	g
1 Skrupel=20 Gran	1.218	"
1 Quintlein=48 Gran	2.923	"
1 Drachma=3 Skrupeln=60 Gran	3.654	"
1 Loth=4 Drachmen=5 Quintlein=240 Gran	14.616	"
1 Unze=2 Loth=8 Drachmen=480 Gran	29.232	"
1 Pfund=12 Unzen=5760 Gran	350.784	"

BIBLIOGRAPHY

C. Baker, Modern Gunsmithing. Marshalton, Del. U. S. A. 1928.

Bauer & Roth, Technologie der Gewehrkunde. Suhl, 1923.

Dr. E. Beutel, Bewährte Arbeitsweisen der Metallfärbung, Wien & Leipzig, 1925.

F. Brandeis, Die Moderne Gewehrfabrikation. Weimar, 1883.

Brannt & Wahl, Techno-chemical Receipt Book. London, 1920.

G. Buchner, Metallfärbung, 6. Aufl. Berlin, 1920.

Colvin & Viall, U. S. Rifles & Machine Guns. New York, 1917.

Field & Bonney, Chemical Colouring of Metals. London, 1925.

W. W. Greener, The Gun & its Development. 1846 to 1910.

A. H. Hiorns, Metal Colouring & Bronzing. London, 1892.

F. Hartmann, Färben der Metalle. Wien, 1912.

J. V. Howe, Modern Gunsmith. New York, 1934.

M. Hufschmidt, Färben der Metalle. Dresden, 1912.

Instructions for Armourers. London, 1885, 1909 & 1912.

Instruction sur les Armes Portatives. Brussels, 1906.

H. Krause, Metallfärbung. Berlin, 1922.

H. Mangeot, Traité du Fusil de Chasse. Brussels, 1854.

J. Michel, Coloration des Métaux. Paris, 1931.

C. W. Sawyer, Our Rifles. Boston, Mass., U. S. A., 1920.

J. Schön, Geschichte der Handfeuerwaffen. Dresden, 1858.

H. Schuberth, Hand- und Hilfsbuch für den Praktischen Metallarbeiter, 2. Aufl. Wien, 1922.

Stelle & Harrison, Gunsmith's Manual. New York, 1883.

F. Voytier, Gun catalogue. St. Etienne, without date.

J. H. Walsh, The Modern Sportsman's Gun & Rifle. London, 1882.

Colonel Townsend Whelen, The American Rifle. New York, 1920.

Periodicals.

"The American Rifleman," formerly "Arms & the Man" and "Shooting & Fishing."

"Metallarbeiter," Wien.

"Metal Industry," New York.

ABBREVIATIONS

anh.	anhydrous.
aq.	added water: with salts, molecules of water of crystallisation.
b.p.	boiling point.
conc.	concentrated.
cryst.	crystallised.
crwh.	crystallised hydrated.
ccm	cubic centimetre (millilitre).
D.	density (specific gravity): when not otherwise stated, at 15°C.
D. 15/15.	density at 15°C. compared to water at 15°C.
D.Ap.V.4.	supplementary volume to German Pharmacopoeia 4th edition.
D.R.P.	German patent.
°C.	degree Centigrade: with alcohol, percentage by volume (degrees Tralles).
°F.	degrees Fahrenheit. (*)
fl.oz.	fluid ounce.
Fr.	French.
G.	German.
grs.	grains (by weight).
g	grams.
hydr.	hydrated.
L.	Latin.
m.p.	melting point.
P.	patent.
pt.	pint.
G-%	percentage by weight.
V-%	percentage by volume.
G/V-%	percentage of weight to volume, that is, grams substance in 100 ccm of solution.
Ph.	Pharmacopoeia:
	U. S. A. American, 8th to 11th revision.
	Belg. Belgian, 3d edition, 1908.
	Brit. British, 1898 and 1914.
	Fr. French, 1920.
	G. German, 3d to 6th edition, 1890-1910.
sol.	solution.
t.s.	(tantum sufficit), as much as needful.

(*) *Note.* All equivalent Fahrenheit temperatures are approximate.

www.ingramcontent.com/pod-product-compliance
Ingram Content Group UK Ltd.
Pitfield, Milton Keynes, MK11 3LW, UK
UKHW041820150726
7214IPUK00001B/1